A USER'S GUIDE
THE SEQUEL

A USER'S GUIDE THE SEQUEL

The Further Adventures of Religion and Science

HERB GRUNING

Foreword by Paul Van Arragon

WIPF & STOCK · Eugene, Oregon

A USER'S GUIDE—THE SEQUEL
The Further Adventures of Religion and Science

Copyright © 2022 Herb Gruning. All rights reserved. Except for brief quotations in critical publications or reviews, no part of this book may be reproduced in any manner without prior written permission from the publisher. Write: Permissions, Wipf and Stock Publishers, 199 W. 8th Ave., Suite 3, Eugene, OR 97401.

Wipf & Stock
An Imprint of Wipf and Stock Publishers
199 W. 8th Ave., Suite 3
Eugene, OR 97401

www.wipfandstock.com

PAPERBACK ISBN: 978-1-6667-4238-1
HARDCOVER ISBN: 978-1-6667-4239-8
EBOOK ISBN: 978-1-6667-4240-4

08/26/22

Unless otherwise indicated, all Scripture quotations are taken from the Holy Bible, New International Version®, NIV®. Copyright © 1973, 1978, 1984, 2011 by Biblica, Inc.™ Used by permission of Zondervan. All rights reserved worldwide. www.zondervan.comThe "NIV" and "New International Version" are trademarks registered in the United States Patent and Trademark Office by Biblica, Inc.™

Scripture quotations marked (NRSV) are from the New Revised Standard Version Bible, copyright © 1989 the Division of Christian Education of the National Council of the Churches of Christ in the United States of America. Used by permission. All rights reserved.

For my wife, Alice,
Sine qua non

CONTENTS

Foreword by Paul Van Arragon	ix
Preface	xiii
Introducing	xv
Objections	1
Channel Three	6

PART 1: RELIGION

The Bible: Edited Version	37
What's the Use?	44
Who's on First?	46
Addenda	49
Front Door Blues	54
As for Myself	58
Which Lives Matter?	60
Pride and Prejudice	65
The Doors of Perspective	70
"If You Build It," They Might Stay Away	75

PART 2: SCIENCE

The Doctor is Out, and so is the Jury	81
The Seventy Percent (Range) Solution	85
Exigency	88
What Will We Think of Next?	93
Visitation or Desperation?	103

PART 3: RELIGION AND SCIENCE

Quantum Conundrum and Other Riddles of Science and Religion	115
Physics and Philosophy	115
Newton	116

Theology 1	119
Relativity	121
Theology 2	124
Quantum	126
Theology 3	134
Others	137
Is it Time to Panic Yet?	142
Concluding	145
Appendix	153
Bibliography	155

FOREWORD

Ever since I was a teenager, I've always wished there were a user's guide to help me understand the world. And suddenly, decades later, there it was, in print for all to read. By my friend, Herb Gruning.

And now there is even a sequel. I guess it is not surprising that there would be a sequel. How could one express all one needs to know in just one volume? After all, as Gruning himself says, "the world continues to be a messy place and will not conform to a neat and tidy equation."

Well, neither does this sequel.

Gruning picks up right where he left off by dealing with objections to his previous ideas. If you haven't read the original guide yet, and you are puzzled about these objections, fear not! Whether the sports stars of today are as amazing as the legends of our youth, and how that relates to science and religion, doesn't ultimately change much in the larger scheme of reality. As they say, it's the thought that counts. Or, more to the point, it's the way of thinking that counts. If you reflect upon Gruning's writing, you will find fair-mindedness, attention paid to curious details, willingness to ferret out the flaws in one's own tradition, and openness as to where it all leads.

Once the objections have all been parried, Gruning treats us to a wonderful fictional dialogue, that could stand on its own. Although written as fiction, it elucidates the real-world interaction of ideas from diverse sources at play in Gruning's mind. Literature, paleoanthropology, and physics, all communicate with each other in an imagined discussion using rhymed couplets. The characters are Thomas Mann, Teilhard de Chardin, and Albert Einstein, who analyze each other's legacies. They respect each other, but become critical, almost insulting, with each other as well. They all get credit (and blame) for how Gruning thinks today.

What follows, in the religion segment of the book, continues this approach of critical respect, as Gruning explores the ancient traditions in the Bible. He respects the biblical authors' attempts at understanding the world, and agrees with their basic premise that a higher power is at play, and yet he ruthlessly subjects them to scrutiny, pointing out their many inconsistencies. It is as if there can be no place for these traditions in our modern scientific world unless we give them a good shake to see what survives.

This approach reveals that the description of God in the Bible isn't always consistent, and often it describes a higher power we don't want to believe in: one who promotes genocide, for example. And, as Gruning points out, there isn't steady progress in the Bible towards understanding God as a God of love. In some cases, the New Testament God is less forgiving than the Old Testament God.

There is a good deal of pleasure in the journey as well, partly because of the novelty of thought and the humor. You just never know what you are going to find when you turn a page. Front door evangelists, parapsychology, climate change, racism, objectivity, ancient medicine, pareidolia, apophenia, synesthesia, and whether aliens helped build the pyramids are just a few of the subjects under consideration.

The intent is always serious, but the mode is playful and creative. For example, it would not have occurred to me to attempt to overcome the biased agendas of the biblical authors by focusing on their parenthetical statements. But, as Gruning explains, parenthetical asides are less likely to be biased, because they tend to reference agreed upon facts, which are not in themselves agenda-driven.

Gruning, of course, has his own agenda. Religious discourse in our time is suffering. Many wonder whether religion can survive this century as anything more than a mechanism to control people or to justify atrocities. And yet so many people still rely on these ancient traditions to orient themselves in community and to propel these communities towards love and positive action and healing the world.

Likewise, science has many detractors in our modern world, and could benefit from a credibility boost, particularly among religious people. We must pay attention to the science about the human causes of climate change, for example. Although Gruning is critical of science for, at times, overstepping its boundaries as if empiricism is the only way to knowledge and as if its own assumptions aren't presumptuous, in this volume, the emphasis is also on encouraging people to learn from science, and to heed its warnings.

Gruning also explores how science and theology inform each other, by explaining some of the key ideas of relativity theory and quantum mechanics, and how they affect theological notions of divinity. In keeping with the beginning of the book, where Gruning describes how diverse sources formed his own thinking, here we see how new scientific ideas have theological implications.

Overall, Gruning makes a good attempt at enticing people out of their self-imposed confines from, as he puts it, "the either-or camp into a both-and approach", but not at the cost of confused or lazy thinking.

However, as a user's guide to our present world, one would need a much larger volume, and even so one would not be able to explain everything. We can always hope that some day there will be a sequel to this sequel.

Paul Van Arragon

PREFACE

A former colleague of mine from a different department, namely English, once fielded a question from a student at a conference setting who asked him what we are to make of religion nowadays. His response was that we should all grow up, intending by this, my knowing him, to mean that we should dispense with it. A common sentiment. Well there are many of us who actually have matured and yet retained our religious affiliation. On the one hand, something happened in history that has import and implications for all of history. We find ourselves attracted to Christianity precisely because of the realization that we cannot meet God's standards for our lives through our own effort and thereby choose to follow the only one who could and did. He now has our attention. And on the other, this person has a claim on all of life, probably the lone ideological holdover from my wife's and my Reformed upbringing, and that includes the subject matter before you, specifically the studies of religion, science, and the two combined. Consequently, the fact that this strategy drives me to be critical of all three approaches should not be interpreted by the reader as my deprecating these disciplines, rather my seeking that they be refined. This disclaimer is offered as an alert, for there are those who might fail to make the connection. At least now you know what to expect in another installment below.

INTRODUCING

Here we go again! I mentioned in my previous volume that it was a sequel to the one before it; so here I am with similar intent in this title—a sequel to a sequel. I am not certain as to whether this is even good form. Oh well, let's chance it. Why mess with a good thing?

The earlier monograph marks a radical departure from my usual approach of preparing manuscripts containing a sustained argument, in that the newer versions are highly episodic, covering many, often unconnected, themes. The tendency is such that when investigators' studies conduct their commerce in a popular vein, sometimes precision is compromised, though this can be remedied by reference to other sources, some of which include my own. Admittedly, the charge could be laid that the former work was a series of editorial reflections cobbled together and geared toward ranting against what I regard as inappropriate ventures into religion, science, or both. There might be some truth in this, but not entirely. I remain an equal opportunity critic and seek to repair the broken. What I am about here is following up on this tradition with another set of installments. I take as my model the tradition of Jesus as master storyteller, a provocative raconteur, who employed story and parable to spark discussion. If his listeners were not careful, they would find themselves the very targets of the tale once they arrived at the other end. The method turned out to be effective. Jesus serves as an exemplar we can follow for this task.

Prior to my academic career, I thought I would never have anything to say, that is, to add to what has already been contributed by others in my field; now I cannot seem to refrain (I am prevented from writing "shut up" for running the risk of being indelicate, although, come to think of it, this has not stopped me before). I have employed a combination of the above biblical parabolic strategy together with the Socratic approach in my own teaching, giving my students the tools and being a resource person, leaving them compelled to embark on a voyage and come to conclusions on their own, recognizing both the strengths and weaknesses in both theirs and all other positions. I am there as a guide to steer discussion and act as referee when we digress. But we are not here to converse on philosophies of education; instead, in order to accomplish the mission of keeping the conversation alive, we pursue the end of asking more informed questions and considering potential options. In our quest we might even profit from the exercise by becoming enriched.

OBJECTIONS

In an effort to demonstrate that I engage with prospective critics, I commence with the following objections, reversing the trend only in this section from previously where religious topics were covered first and then scientific, since in this case the former is the longer. Anticipating an objection from last time, when I made a critical assessment of Stephen Jay Gould's theory of the reduction of excellence in human abilities in his work *Full House*, with specific reference to baseball, another study has put my view into question. Admittedly, experimentation has revealed that were stellar athletes of today to don the gear of yesteryear, such as footwear, and compete on the quality of surfaces for, say, sprinting, then they might not even reach the levels which previous athletes were able to.[1] Here are some comments.

First, were athletes of today to practice in and with these conditions for extended periods, instead of being subject to them but once, they could very well reach if not surpass the marks of yesterday. This could actually be attempted. What could also be examined is how current athletes would fare had they been exposed to the nutritional, coaching, and training regimen which athletes encountered in years gone by, together with providing them with spectators and opposing competitors. Second, an experiment which could not be performed, regrettably and obviously, should they already have passed away, is whether yesterday's athletes, given sufficient time and opportunity, could ultimately equal or even eclipse today's sporting marks using the resources of the present, so we cannot be certain. Lacking this information prevents us from being definitive about comparing yesterday's and today's sports stars, especially if the record-setting athletes used as subjects in the study are significantly removed in time from those events, which some were.

Third, there are some aspects of sports which remain the same throughout the ages. For instance, those football kickers who can now kick field goals, with some regularity, at or past the sixty-yard stripe are not using exceedingly different footwear from before nor does the ball itself vary tremendously, thereby enabling kickers to reach those standards more easily; the changes are not enough to make up the difference when it comes to record setting and breaking. There are no asterisks in the statistics columns

---

1. Suzuki, "Equalizer."

stating that inferior equipment was employed at the time. In fact, the object of play in the various team sports, such as baseballs and hockey pucks are much the same as before, though gloves, bats, sticks, skates, and protective equipment have indeed undergone significant modification. Basketballs, though, are different from the early days, but this does not amount to revolutionizing the game by making them markedly easier to handle and shoot. These are the criticisms I have concerning the recent investigation. The upshot of all this is that the present author need not retract a great deal of the former piece as yet, nor is the research into comparing the performance of athletes from different eras rock solid.

Next, as a point of clarification concerning the topic of cosmology, the mathematical model of the end of the universe among the possible ones is actually known, based on studies involving the cosmological constant published in articles from January and March of 1998, which indicate that we are currently in a flat universe which will become an open one owing to the runaway acceleration of the universe's expansion rate[2] (this will be elaborated upon near the end of the study). What is not clear, however, and which throws the above into question, is that no model could have foreseen the expansion rate as having inexplicably begun to accelerate about five billion years ago,[3] nor is there a mathematical model which can predict if and when this acceleration might do something else inexplicable, such as decelerate. If it did so once, we must concede, it could do so again. If we are honest, we will admit, as mentioned last time, that we do not know how universes behave long term, hence mathematical models by themselves will not be conclusive.

The biblical objection, in a further attempt to whet our appetites, calls for some background. Prayer is an integral part of the Christian life. *That* we should pray is beyond dispute; *what* we should pray about and for, outside of the Lord's prayer (Matt 6:9–13; Luke 11:2–4), is not as straightforward. When we are given instructions in the text as to how to act, sometimes the direct opposite occurs in the very next verse. In Prov 26:4–5, for example, we are enjoined not to "answer a fool according to his folly, or you will be like him yourself," and in the immediately following verse, we are exhorted to "answer a fool according to his folly, or he will be wise in his own eyes." The manner in which we respond to these two verses betrays what type of tradition informs us: the more recent in time is the Greek philosophical, which elevates rationality and logic. For them, the reaction to these side-by-side verses would likely be, in contemporary computer terms (think

---

2. Wilford, "New Data," "Wary Astronomers."
3. Randall, *Dark*, 62.

monotone), "this violates the law of non-contradiction and does not compute." The earlier Hebraic mindset is not so troubled with the potentially confusing juxtaposition because what it elevates is wisdom: it requires wisdom to unravel and disentangle as to when each situation is warranted and God's Spirit is present to guide us along the appropriate path. This is one way of showcasing that a Hebrew divinity does not fit into a Greek mold so easily, and thereby the task of rendering theology in a systematic way is a Greek innovation and the fruit of a Greek perspective, not a Hebraic—one reason why the theological task requires continual reformulation and is never final and complete.

For the prayer theme, I decided to undertake a word study, not on prayer but on the term "ask," that is, what it is we are to ask for when we pray. Interestingly, the word is not even listed in the short concordance placed at the end of my NIV, so I needed to consult the heavy-duty Strong's Exhaustive Concordance, which contains a lengthy entry on it, and requires its namesake in order to negotiate (would that his surname had been "Weakling").

The pertinent passages are the following, beginning with the Old Testament (OT) and moving on to the New Testament (NT). The initial verses for our purposes are 2 Sam 12:7–9, where God was incensed at David's adultery and, like a taxed parent, through the prophet Nathan, chided that God had given David what he wanted: particularly the kingship in place of Saul, Saul's house and wives, and even all of Israel and Judah, and should all of this have been insufficient, God would have rendered still more. The asking is implied here, but it seems the requests must be in line with God's commands. Nevertheless, David did not actually petition God with his requests as such, hence the asking does not appear to be a requirement for obtaining what one seeks.

The next concerns David's son Solomon, who became king subsequent to David. God invited Solomon to ask for whatever he wished; he did so and was bestowed with wisdom and a discerning heart to rule over the people. Not only was he granted his request, but since he did so from a humble heart, he was also given what he had not asked for, specifically riches, honor, and potentially a long life, to be extended were he to keep God's commandments and follow in the ways of his father (minus the adultery, of course) (1 Kgs 3:5–14). Thus in both instances, these figures were granted what they had not asked for, evidently making the asking of secondary importance.

In the NT we are informed that God knows our needs prior to our asking (Matt 6:8), but soon afterwards we are instructed to ask, seek, and knock, as then what we ask for will be granted (Matt 7:7–11; Luke 11:9–13, although the latter passage alters the object of what is to be given from "good gifts" to "the Holy Spirit"). Later in Matthew (18:19–20), we are told plainly that if any two or three who are gathered together in the Lord will ask in his name for whatever they seek, it will be granted. In John's gospel, we may ask Jesus for anything in his name and he will perform it (14:13–14; 15:7, 16), and since we have not as yet asked, we should (16:23–24).

For all of these passages the specific request comes without qualification, hence it could virtually be anything. It is not until the writings of James and later 1 John that the petitions are conditional. For one, the asking is re-emphasized, though it must come from a place of proper motives, not for our personal gain (Jas 4:2–3). For the other, in 1 John, God hears us whenever we ask God for anything, yet it must be in accordance with his will (5:14–15), and only if we are obedient to God's commands and follow God's will, walk in God's paths, and focus on God's pleasures, not our own (3:22). Perhaps the "anything" in the previous verses needed greater precision should the requests have gotten out of hand, necessitating the qualifiers in the latter ones. Furthermore, the unasked for gifts in the OT, which were nevertheless granted, in the NT come with the requirement to submit or make supplication for what we pursue.

In a related episode, Hezekiah king of Judah developed a fatal illness and was informed by Isaiah, a prophet of God, in no uncertain terms that this was to mark his demise, stated flatly by God. But after many prayers and tears on Hezekiah's part, God relented and added fifteen years to his life. Hezekiah even asked Isaiah what the sign to him would be that this was about to occur. He was not reprimanded for any insolence, but, by his own choice, a shadow went backward ten steps on a staircase (2 Kgs 20:1–11). Aside from the obvious astronomical infraction, this again differs from the NT, where in Luke 1 Zechariah the priest and father of John the Baptizer was foretold by the angel Gabriel about the birth of his son. He asked how this could be confirmed given his age—in essence asking for a sign. Due to his disbelief he was struck dumb by Gabriel until eight days after John's birth. It has sometimes been stated that the God of the OT was sterner than in the NT; occasionally the reverse is true.

The reason I broach this topic is the following: the enigmatic formula seems clear to us now. Plainly, in a Greek fashion, if we do A in a B way, then we will obtain C. Regrettably, it is not so simple, for certain unscrupulous televangelists have used this to their own advantage and capitalized on our vulnerability. They are oriented to marketing themselves as more effective conduits for success in these endeavors, since whether through lack of faith or not asking in the right way, we are solicited to kindly donate to their program and they will do the asking for us. In essence, they are offering an opportunity to purchase an answer to prayer, much like what the church promised at the time of Luther and which propelled him to strenuously and vociferously object to its use. Ostensibly, the contemporary practice also requires reformation, and we need not be made to feel like we are inferior caliber Christians for not securing what we plead for on our own. We are not always served well by those who adopt the same label.

The world continues to be a messy place and will not conform to a neat and tidy equation, despite the promise of it in the NT. The main promise that we can invest in is that God will continue to bring order out of chaos and redress grievances when all is said and done and history is drawn to a close. The cost of discipleship is a more accurate assessment and take away from the wider biblical message. Even the apostle Paul feverishly sought God to remove his infirmity, God instead bestowing grace upon Paul with which to bear it. Consequently, there is at least one major instance in which God was asked for something but did not grant it, nor was the request made from improper motives. And Paul was no second-rate Christian. As a result, we are to live with the scriptural mixed message in a healthy tension. Does this undermine the scriptures? No, it just turns the idealistic view in a more realistic direction. Similar to the Proverbs passage, this defies logic and calls for wisdom.

And lastly, commenting on terminology, it is inaccurate that all statements containing the terms "nothing but" are reductionistic, for one might be so inclined as to declare, "nothing but California wines for me, please," and in so doing one is not being a reductionist, for other wines have their place even if not for oneself.

Having completed these introductory remarks, I now wish to shift, similar to last time, to a work of fiction, this time a somewhat lengthy short story (one of my favorite oxymorons), so as to broach some of the topics to be amplified in due course.

CHANNEL THREE

Once upon a time there were three men. These three gentlemen are not to be confused with the subjects of some inane, banal anecdote commencing with the line "There were these three guys, see," and often combined with their whereabouts "in a bar." These fellows were contemporaries in both time and space. But of course I speak as a Westerner in the New World. Recall the old tale of a traveller from the British Isles to North America who, when asked to contrast the two worlds declared, "In Britain, one hundred miles is a long way; in North America, one hundred years is a long time." From the perspective of a European (aside from the British notion that Britain is not properly in Europe, since not on the continent), then, the distance separating these three was not inconsiderable; from the perception of the New World, however, they were next door neighbors. For the latter, to speak otherwise would be to make a mockery of scale.

I mention space and time since one of the conspecifics was Albert Einstein, who heralded in a new age of combining the two. His was a revolutionary way to view the universe and our place in it, and we have never been the same since. But in eagerness I am getting ahead of myself.

There was also a fourth man, who came later and was thereby not a contemporary of the other three. For reasons of privacy and by his own request, he prefers to remain anonymous and so shall be rendered nameless, though by all accounts he did bear one, in fact three, as is typical of his culture. In an attempt to describe him we refer and defer to the remaining two of the initial three. Pierre Teilhard de Chardin, a noted paleoanthropologist, a fancy term for a hunter of early human remains, would uncontroversially confirm that our fourth figure is a descendent of archaic *Homo sapiens* and has a common ancestor with *Homo erectus* and other extinct human species. Further, also uncontroversially, many modern *Homo sapiens* contain anywhere from one to four percent genetic markers of *Neanderthals* in their genome,[1] indicating there to be an overlap between the two and hence making the two—how should I put it delicately?—conjugals to an extent, and by virtue of producing fertile offspring were not as yet different species but were, so the debate runs, subspecies at least for several generations. This

---

1. Fagan, *Cro-Magnon*, x.

latter discussion was not one to which Teilhard knowingly contributed, since it was left to subsequent generations of researchers, particularly geneticists and biochemists.

The last but definitely not the least of the trio was a literary personage named Thomas Mann. He along with our fourth shared the sometimes ignominious distinction of being of German heritage, Mann himself along with the Hebraic Einstein understandably fleeing Third Reich Germany and making Princeton University in New Jersey their professional destination. Sigmund Freud, not one of our triad, was also a Jewish refugee, this time from Vienna, but fled to England in 1938, only to pass away three weeks into World War II, on September 23, 1939.

To this point, describing the fourth man as a German *Homo sapien* fails to appreciably narrow down his identity—exactly the way he wants it. We can, nevertheless, be more precise in terms of his proximity in time to the three. All three of them experienced their demise in the year of our Lord 1955. Our fourth was born in 1958. For the reincarnation aficionados among us, the intervening three years marks the conventional time limit between the perishing of one person and his or her rebirth into another body, should what remains of the first person cross over or transmigrate into the next in line. But we are not speaking of reincarnation *per se* here, since three persons becoming transferred into one is a transaction in apparent contravention of the reincarnation operations manual. Nevertheless, as for his birth, our fourth man was delivered on a Sunday at 2 p.m., continually casting him in the role of an afternoon, evening, and night person, his waking day having been shifted from the norm by six hours ever since his arrival on Earth. But as is typical of his species, he objected to the rude intrusion of his delivery by weeping and wailing.

Yet we can catch a glimpse of what our fourth is like with the writings he has left behind. Below is an example of a Christmas bulk mailing he composed, prepared but never posted, at the end of the year 2020, when it was supposed that this year would be the solitary one in which the world would be held captive by the new virus.

> Our thanks [written on the eve of he and his wife's thirty-fifth wedding anniversary] to all those who have corresponded with us as to this year's events in their lives. Now it's our turn. We are amazed as to the optimism of these reports, both in the way of personal experiences amidst a pandemic together with anticipated prospects for the future. Or at least we are putting a positive spin on a potentially dire situation. We too are grateful for the mercies which have assisted us in evading the rapidly

mutating virus, though it has touched some who are dear if not also physically near to us.

The trouble at times is that if one is not in close proximity to those cases which have tested positively and are presenting symptoms, then this tends to encourage others who are anti-pandemic and/or anti-vaccers in their anti-alarmist message and practises, while employing precisely an alarmist strategy in their attempt to convey it. Insisting on personal freedom can result in compromising their own as well as others' safety, while they themselves would also expect to be treated, should they come down with the condition, by the very health system they are jeopardizing. Meanwhile, to the neglect of other patients, many die because they cannot obtain the surgeries they require so as to preserve their lives, owing to the selfishness of those unvaccinated who occupy the beds in hospital wards and intensive care units.

Moreover, given that these resources, including respiration equipment, are in short supply and essential workers are strained to the brink, health care providers are driven to resort to the psychologically gut-wrenching policy of triage, which, when facing multiple patients with dire prospects, are then called upon to make a snap decision, sometimes by taking a purely random selection process as to which one gets to receive treatment while those left out are cast adrift toward a certain death, leaving the staff, though not the loved ones, who are not allowed to visit, to witness the agonizing demise play out. Perversely, freedom for some, evidently for them the most sacrosanct ideal on the planet, or second only to family, means loss of life for others. With all the strength I can possibly summon, together with all the restraint I can possibly muster, what I can say to these folks about their ostensible self-centeredness is, "Oh, careful, your ignorance is showing!"

As for us, we have dutifully isolated when called upon to do so, nor has this produced much in the way of hardship. Hubby in particular has encountered little in the way of lifestyle alteration; now he just has an excuse for it. Having said this, the requirement of educating via Zoom has lost its novelty and we suffer more from Zoom-fatigue than Covid-. Plus, we have not been able, customarily, to travel to Florida this year and are experiencing withdrawal symptoms as a result. After ten consecutive years of frost-free Christmases, the eleventh is white (a dream we do not countenance). Should this be the worst our lives become, then they are more than sufferable. We wonder, though, if we would have the same perspective if the circumstances were to be protracted into next year at this time and perhaps even further down the line.

We must say that we believe Hallmark cards to have really missed the mark when it comes to the coronavirus in not having seized upon the opportunity to craft such greetings as "Merry Quarantine," "Happy Self-isolation," "Blessed Vaccination," "Lucky Lockdown," and more recently "Break-a-leg Booster." Regardless, ours is a relatively mild adversity, but we do hope that the restrictions will be lifted for us to resume with all of you in a more tangible rather than a solely electronically-assisted personal contact.

Given the above, what we can disclose is that the fourth man hails from a northern clime for which Florida is held to be a substantial seasonal remedy. Actually, by all accounts he despises winter and speaks of it disparagingly. He tolerated it as a child and put on a brave face when playing road/street hockey, but he could abide it no longer, believing he had paid his dues. In fact, if the seasons were to be ranked on a scale with one being the highest, for him winter would sneak in at number five. (Indigenous summer, so as to be politically correct, rates much higher.) Plus, he seems to accept the pandemic protocols put in place (note the use of a quartet of terms beginning with "p," something about which he would also give his approbation) by local and federal governments.

Back to the triad. Research has uncovered additional points of similarity as well as dissimilarity between the three and our fourth. Please be aware that the document you have before you is a work of (hopefully) creative nonfiction sprinkled with elements of speculative fiction. Sometimes it is unclear as to which portions belong in which category, for there might be no other recourse. So kindly hear the following conjecture and decide for yourself.

But I digress. As for Einstein, who was born in Ulm, Germany, our fourth has relations in the nearby town of Biberach. Both comment on religion and science and both have taught at the tertiary level in the U.S., though the fourth has spent most of his vocation as an educator across the border in Canada. The fourth is not Jewish but mainline Christian and is more demonstrably German in the sense that he imposes greater order on his environment, in distinctly German fashion, than Albert, one example being that our fourth does not wear his hair as though he has stuck his finger in a wall socket and, not necessarily a German quirk, he abhors the pastime of sailing on a boat for pleasure.

When it comes to Teilhard, the fourth is not Catholic though has taught at a Jesuit college in Buffalo, New York, named Canisius. Both have an interest in paleoanthropology and both are enthusiastic about relating, even integrating, their religion with their science, particularly in terms of what this tells us about the nature of God. While both have seminary training, only Teilhard is ordained. And while Pierre is a Frenchman who hails from Orcines, the closest our fourth comes to French is that he lives in a nation where French is the second official language. Nor, understandably, is Pierre a married man. Despite their differing heritage, our fourth might have the greatest similarity with Teihard due to where their work in religion and science have taken them philosophically and theologically.

Lastly, with respect to Mann, born in Lubeck, both he and our fourth are of German lineage and are interested in literature, though only the former made a living from it in addition to his educational employment. There are moments, few in number, when our fourth has displayed similarity in style before ever having been exposed to Mann's work. He finds this particularly in Mann's short story "A Man and His Dog."[2] They share a native mother-tongue though not a native land and came to English secondarily. Permit me to make the following observations about the two languages with which both of them are most familiar. It has been claimed that if one really wishes to conduct oneself philosophically, one is counselled to do so in either Greek or German, owing to their precision. Note that English is not one of them. This, however, should not cause German-speakers to be presumptuous. Here is what I mean.

Instances of English idiosyncrasies include two of the seasons, namely summer and winter, being employed both as verbs as well as nouns—think of snowbirds, in addition to a unit of measurement, specifically an "inch," used both as a length and the pace one can assume. It is a further peculiarity of the English language that an "affront" is not the opposite of to be "taken aback," nor is "backal" the opposite of "frontal." Wait, there's more. The opposite of a "rental" is not an "ownal," nor is "aclose" the opposite of "afar." The opposite of "shortly" is not "longly," of "utmost" is not "utleast," of "winless" is not "lossless," similarly of "undefeated" or "unconquered" ("invictus") is not "unvictored," and of "belittle" is not "bebig." Moreover, a "repeat" is not a second "peat." "Flammable" and "inflammable" mean the same thing, and seldom do we see a double "o" pronounced as it is in the term "blood." We also observe many terms having unpronounced consonants as in "two," "eight," "castle," "half," "debt," "phlegm," "whale," "who," "wrong," "island," "right," "Isle of Wight," "knight," "knot," and most importantly "know." Finally, the "z" in "prize" for some reason becomes an "s" in "surprise," the "e" in "lie" mystifyingly changes into an "a" in "liar," "past" is pronounced the same as "passed," and "excuse" is pronounced differently whether it is used

---

2. Mann, *Death*, 217–91.

either as a noun or verb, as in "Excuse me, but you have no excuse." As a response to all of this our subject can merely "sigh."

On the other hand, lest German presume it is immune from inconsistency, imagining as it does that every letter in a word is pronounced, there are occasions for example in which the letter "h" is silent as in "Floh" (flea), "Stroh" (straw), "froh" (happy, glad, satisfied), "sehr" (very), "ehrlich" (honest), "fehlen" (lacking), and "Weihnachten" (Christmas). Plus, what would the sound differential be between the German letters "f" and "v" as in "Volksfest" (the people's celebration), entailing that they are interchangeable? Lastly, forms or combinations of terms in German which are unnecessary in English include "Ich wunder mich" (I wonder me) as well as constructions requiring additional terms when translating directly into English, like "das schmeckt" (that tastes, where "good" is implied). Both Mann and our fourth have had to deal with both sets of quirkiness. (And as for a misnomer common to multiple languages is the world's greatest one, namely Greenland.)

Some biographical notes from our research are in order for our fourth man, let's give him the name Wolfgang both to preserve his Teutonic heritage, it was, after all, Mozart's first name, and before it becomes exceedingly tedious to do otherwise. There were some instances in his childhood when Wolfgang withstood odd moments in the classroom. In the winters as an elementary schooler, he would often have trouble with his throat and would seek to clear it, thus making himself audible to the teacher, let's call her Miss Blunderbuss. She would peer at him with eyes from which fiery darts would issue and ask if the noise he was making was necessary. This was asked with a look that would make a prosecuting attorney force a confession out of a defendant. At this moment his philosophical antennae would be piqued and only much later would have responded with, "Why no, it is entirely contingent upon the amount of phlegm lodged therein along with my feeble attempts to expectorate it. Never does it reach the level of the inevitable let alone outright necessity, the former intending that something will occur and the latter that something must occur, for a distinction is warranted here. Frankly, I find it disconcerting, especially for one in your educational position, that you did not work this out."

On another occasion, Wolfgang received back a written report he completed on some subject or other now lost in the mists of time, for which he received a not altogether objectionable grade. A comment from the teacher, let's call her Mrs. Howitzer, was placed in the margin stating "not a word." His reaction was to come face to face with an ambiguity, which struck another philosophical chord in him. Was this meant to convey, he asked himself, that the term in question, now inconsequential, was a fabrication not to be given approval by anyone named Noah Webster, or was it a command similar to "Thou shalt henceforth and hereafter neither write nor utter such a term in these precincts on pain of censure"? He shrugged and simply concluded that the world suffered from short-sightedness and that the users of words should be accorded the latitude to be creative with their meaning and spelling since consensus is overrated. Any takers?

His father early instilled in him a passion for the sciences and fed it by occasionally bringing home a book about it, which Wolfgang absolutely devoured in excitement. This is something his father himself would have pursued back in Germany had he the chance to progress beyond grade school. His mother also facilitated in Wolfgang an interest in Christianity, marking different celebrations and comparing differences between German and North American observances. These latter efforts did not really "take" until secondary school.

On a personal note in terms of taste, Wolfgang prefers the sea and lake sides as opposed to the mountains, having experienced both as far as travels to Europe with his parents and later with his wife are concerned, for he shies away from the thin air of higher elevations but would rather that the air be positively obese. And he also had visions of being the front man in a blues band and, in the tradition of musical artists like Sting and The Edge, renaming himself "Hemorrhage," but he figured it would just wantonly be mispronounced "Hemorrhoid." One more introductory item. Wolfgang is susceptible to colds and often thinks he is coming down with one even when he is not. He feels the same about the cars he drives, that they are in constant need of repair. So I suppose he could be described as a sympathetic mechanical hypochondriac.

In high school, Wolfgang concentrated on the maths and sciences, intending to make his career a medical one. He became fascinated by the subject of astrophysics ever since reading Lincoln Barnett's popular volume *The Universe and Dr. Einstein*, which the librarian there presented to him as a gift, seeing how much it impacted him. He retained this interest despite the presence of, let's call him Mr. Poindexter, in the final physics course in the program. The reason was that Wolfgang was uncertain as to whether this instructor, given his eccentricities, belonged to the fraternity of human, cyborg, or automaton. His interest, though, waned as a freshman at university, despite taking the course from the one who wrote the textbook, neither it nor his delivery in lecture inspiring an awe of the cosmos that Wolfgang craved. Both were sadly underwhelming.

He further retained an interest in biology, a testament to Teilhard's influence, despite his exposure to, let's name him Mr. Peabody, early in high school. It was during this period that Wolfgang came into the thrall of creation science, much to Pierre's consternation (a theme to be explored forthwith), and backed away from additional biological courses until such time as he attended a Christian university in the U.S. where he figured he would be safe. At the risk of skipping well ahead, he kept up this perspective throughout his undergraduate career, and it was not until graduate school in religious studies when he made an about-face and finally accepted evolution as not inimical to religion.

Wolfgang enjoyed the maths and sciences throughout high school, though his greatest sense of accomplishment came from writing essays in English class. They were always on topics of the extramundane, such as "Were Romeo and Juliet star-crossed lovers?" and "The role of the supernatural in Julius Caesar," and he looked with pride upon his finished products and was rewarded in terms of grades. The introduction to Mann came late in high school in German class, though he preferred the more thoroughgoing existentialist bordering on nihilist authors, especially Franz Kafka, for what he perceived as their biting commentary about human nature and the world they lived in. He appreciated their creativity, which is curious to say for nihilism, where nothing is of ultimate value, significance or worth (except perhaps their writings informing us of that fact).

It was not until the latter part of his undergraduate voyage, however, that he bid a fond farewell to the sciences in favor of the philosophical, recognizing that he had an insatiable appetite for it and being amazed that someone could actually get paid for this type of work. So he decided to forego becoming a medical practitioner and replace it with being a philosophical theoretician on the eternal verities, for at this time a theological bent also took hold of him.

What did not release its grip on him, though, was his "scientific creationism." During his undergraduate days, he favored a "theistic evolutionary" approach—a more sophisticated version of creation science, which later gave way to the "Intelligent Design" (ID) strategy. Cracks in their armor became apparent when authors such as Michael Behe announcing ideas, for instance, like biochemical systems appear not in gradual stages but all at once given their complexity, were supplanted when it was revealed within a year after the publication of his work that their emergence was in fact a step-by-step process after all.[3]

Soon the door to taking evolution seriously became opened, he was just not as yet settled as to whether evolution covered the major organismic changes with God filling in the minutiae, or God remains in charge of the major design changes with evolution acting as a minor component

---

3. Miller, *Finding*, 147, 150, 264.

(Darwin's position[4] immediately subsequent to the publication of the first edition of the *Origin of Species* in 1859). Further research has unearthed that he still does not regard sixty-six million years or so, the beginning of the move from a reptilian to a more mammalian regime, as sufficient time to accumulate all the variations necessary for a mole-like mammal at the time of the end of the dinosaurs to evolve into paleontologists. Wolfgang came to Teilhard's writings late in graduate school, but was glad there was a competing process scheme to Alfred North Whitehead's (to be outlined shortly), at least for purposes of keeping each side honest.

On the theme of ID, one of his initial forays into philosophy came in ethics class, taught by the famous Darwinian Michael Ruse, who looked out of place in this subject area but was one of the onerous obligations pressed upon him by the philosophy department. It was an evening class, and on one occasion a student came in late, apologizing that supper had gone overtime. This ignited Ruse's ire and he blurted out with, "Ah, belly before brains, eh?" in true British fashion. During this time Ruse became an expert witness at the Arkansas trial on creation versus evolution; the "scientific" creationist stance being that their view should be taught alongside evolution in the school system. Their case was rightfully not upheld.[5] At the Scopes "monkey trial" in Dayton, Tennessee in 1925, one could argue for preventing evolutionary ideas being taught in the schools and expect to win; here the best that one in the creationist camp could hope for was equal time, and even this tactic failed to achieve its goal. Some things do improve with time; some people having come to their senses.

Upon his final convocation, the time came for Wolfgang to apply for a professorial post. One such unsuccessful attempt stuck in his memory. At one institution where he went for an interview, he was judged as too philosophical for the position in religious studies. Regrettable, he thought, for his sub-discipline area was the philosophy of religion, entirely appropriate, he surmised, for the task at hand. He managed, in any event, to be placed on the short list but was ultimately passed over for a candidate who came with training in acupuncture. Wolfgang chuckled to himself, with as much sarcasm as he could possibly marshal, "I don't know how many times my pappy told me, as all pappies should, "son, if you want to get ahead in life you need to have a working knowledge of and experience in acupuncture," but I would not listen because, oh, I knew better!" The episode served as a source of amusement for him ever since.

---

4. Barbour, *When Science*, 9–10.
5. Gilkey, *Creationism*, 266–95.

Nevertheless, before long Wolfgang embarked on a teaching career. His life as an educator would be marred, gratefully, by just a single incident of administrative animosity, but nothing of seismic proportions, mind you. During the period when he was delivering courses at Canisius, he was also under contract to offer a class on religion and science, his most well-travelled course, at a Canadian university. At the latter, the academic dean, let's call her Mrs. Axe-to-Grind, phoned him to issue a threat. Perhaps she had either a friend of the family or a blood relative in his class who had submitted a late paper and insisted that it be graded as though it were on time, something about which his two teaching assistants had not as yet informed him. He had the misfortune of being in the dean's line of fire, who promised to exact revenge against him, should he not comply, in the form of not allowing him to teach at this institution ever again.

Concurrently, the chair of the religious studies department had mentioned to him that the course was no longer sustainable anyway, since the number of social science students in it was insufficient for the department to receive funding for it. This propelled Wolfgang to feel like retorting to the dean, "How would I know the difference?" but he refrained. Unfortunately, the losers would be the large number of students themselves, enrolling as they did from across the campus, from engineers to English majors. The class up to that time continued to attract representatives from most disciplines throughout the time he taught it, but now they could no longer benefit from it. For many, this was their only exposure to religious studies, coming in thinking it was going to be fun watching a non-scientist squirm, and leaving with the impression that religious studies is not so bad after all. Having a background in both science and religion-philosophy, Wolfgang never squirmed, but developed a fine rapport with most class members. He ultimately lamented the demise of a popular course, but then thought to himself, "Oh well, at least the time allotted to commuting will be reduced." None of these instances inflicted irreparable harm, nor did even the cumulative effect, but they were nevertheless indelibly imprinted upon his memory (and for which perhaps there is a term in psychopathology?).

But prior to all this, the trio passed away and were taken through a long tunnel into a blue light (all previous accounts were in black and white, this one in color). They were met by a greeter, not like the ones met up with at Walmart, though, but one who shattered their preconceptions, for he (She? It?) had neither wings nor wore a flowing robe. He was tall, looked to be in what to us would be his early thirties, if the comparison can even be made, and his gender, if he in fact possessed one, was not entirely obvious. As for their location, the four of them stood not on a cloud but in a field, like a meadow, though thankfully free from ticks. He carried something like a document and had a very pleasant demeanor. There were also others like him, some walking, some hovering in mid . . . "air?" The document was not so much a record of events as a strategy for increasing the likelihood of events transpiring. More on this presently.

In order to make sense of what they encountered there, let's examine the descriptions on offer from the biblical book of Revelation, the experiences of the triad likely paralleling, at least in part, with what is presented there. An issue we can rally our efforts around and which will be of interest particularly to Einstein, might be the extent to which the customary natural laws we are familiar with also obtain in the hereafter, or whether the slate is wiped clean and an altogether different set takes their place. Compounded with this is the related question as to whether we will have the same senses we currently enjoy or if they will be enhanced or even augmented by additional ones?

John of Patmos, the stated author of the work (1:9), not surprisingly writes from the perspective of an ordinary human who comes armed with the standard ocular, auditory, olfactory, tactile and gustatory faculties. Judging from his depiction, all of these same senses are in play in the afterlife, save for the olfactory, or at least it is not mentioned. There are obvious sights and sounds, and touch appears to be retained since objects are handled, such as books are opened (20:12) and scrolls unrolled (10:2). What would make this situation different is if objects are affected by psychokinesis, where thought or intention becomes the cause of the effect. Furthermore, partaking of the fruit of the tree of life seemingly will come with a taste, blandness not being something worthy of the hereafter. That smell is not included might simply be the result of it not being integral to John's account, he not feeling the need to be exhaustive (nor indeed, I suspect, could he be). Taking the writings at face value for the moment, perhaps the smoke from the burning of the lake of fire (20:10), a favorite passage of fire and brimstone preachers, could carry with it an odor, though it is unclear as to whether it would reach the heights of heaven.

On the topic of physical laws, it could very well be the case that nothing in that state of existence is properly natural or physical, yet it is difficult for us to envision circumstances in which, say, laws of thermodynamics and the conservation of matter and energy were no longer to hold. Nor is it straightforward to conceive of a state of affairs devoid of gravity. Admittedly, this might simply be a comment on our lack of imagination. But if God sits on an actual throne, for instance, then presumably God need not wear seat belts in order to remain in place. Moreover, entropy (a measure of disorder) might not apply if the inhabitants of the heavenly realm are everlasting.

One aspect of natural law that we could most assuredly do without is the Pauli exclusion principle (or the PEP for short), after Wolfgang Pauli (a different Wolfgang). The principle boldly asserts that no two particles can occupy the same space at the same time. Imagine a world bereft of it. Many of the incidents and mishaps about which we complain while in our mortal coil, ranging from stubbed toes to automobile collisions, could very well be dispensed with in a plane where all bodies are spiritual in nature. Free at last. A contemporary analogue could be dark matter, which interacts only gravitationally, meaning entities "can pass right through each other."[6] Upon reflection, however, there are advantages to life with it, for we could at least still participate in billiards or pool and most importantly embrace a life partner. Contact sports would continue to be possible as would the feeling of the warmth of the sun on our faces and the wind at our backs. Come to think of it, perhaps we had better enjoy life with the PEP while we still can.

What does appear contrary to expectations includes the following. If there is no longer to be any sea (21:1), implying that sea creatures will have no part to play in the new creation, to which I would object were I to be an intelligent octopus or dolphin, then where does the water from the spring (21:6) and river of life (22:1) empty into, or is it continually recycled as in a perpetual motion diagram? Were this water to flow "down the middle of the great street of the city" (22:2) (as in the NIV, whereas the NRSV has "through" in place of "down"), this ordinarily connotes that the city or at least the street itself has a slope. Plus, the gates of the city of the New Jerusalem will never be shut (21:25), but for there to be authentic gates at all, they must have the effect of keeping some in and some out. If resurrection bodies will be like Jesus' own, having the ability to pass through locked doors (John 20:26), and if heavenly doors are analogous to standard ones, then this capacity would defeat the purpose of gates. Plus, the non-inhabitants are ambiguously stated as being thrown into the lake of fire (20:15), whereas in 22:15 they are merely outside the gates. This implies that gates do not serve the usual function, which perhaps even precludes the need for them in the conventional sense. If they are always open, then sentries of some sort must be posted there in order to keep the outsiders out, in which case gates would be "redundantly superfluous."

As mentioned, our trio met with a being who greeted them upon their arrival.

Host: Welcome, I will be your host for these proceedings.

Mann: What is this about?

Host: You have never heard of the "heavenly host" (Luke 2:13)?

---

6. Primack, *View*, 118.

Mann: No, I mean which proceedings?

Host: I am glad you asked. My colleagues and I are here to answer your questions and to help you along.

Mann: Do you and they have names and, besides, along to what?

Host: See, we knew you would have questions, everyone does. The "along" is to see how you can assist in the events of the world history you have left behind. As for names, we are all called "Host" to simplify things.

Mann: How does that simplify when we cannot distinguish you by name?

Host: Well, there is a distinction, it is just inaudible to you. There are voice inflections which you can neither discern nor duplicate. We have no trouble distinguishing one from another.

Mann: What would simplify things for us is for all of you to have different designations.

Host: What did you expect, for us to have names like "Cedric" or "Clyde"? Or do I look more like an "Elwood"? And we do have different designations in that each of us has a different portfolio. That is the main difference. If you wish to address one of us, you can mention what each is in charge of, like Terrestrial Policy Host or Secretary of the Interior Host, for like yourselves we have both an interior and an exterior. Please pardon me for anticipating that question.

Mann: That's a lot to keep straight, and I am not getting any help from my colleagues here.

Einstein: No need for us to jump in, you are doing fine.

Teilhard: Ditto.

There was no lack of the newly dead. It works on the basis of first in, first out, so you must wait your turn. This policy means that the newcomers must be supplied with some type of accommodation, not like a health professional's waiting room, please understand, for such an environment is much too sterile, not in the biological sense of a surface with no organisms, but as a place where you would not like to linger in for protracted periods. Face it, it's not a great place to meet people, with nary an eye contact, lousy outdated magazines, and a receptionist or administrative assistant ("don't call me a secretary"), who fails to exude effortless effervescence (pardon the alliteration). So while the recently deceased are in a holding pattern, they get to enjoy a meet-and-greet with others who are also queuing up for new instructions. It's a time to chill with homies and others, but do not lose your place in line and keep up the pace.

First you are outfitted with a blue frock to wear, not a purple one—that's the color of royalty, and not a black robe—too many bad experiences with French priests in the New World, and definitely not white, for only God gets to hand these out (Rev 3:4–5; 6:11). You don't want to be found naked even in the hereafter (2 Cor 5:2–4). These are not permanent attire, so kindly be considerate for the next person. They are also one size fits all, so do hold off on complaints about their not being sufficiently flattering. The permanent ones come later, when all is said and done.

The Hosts are here to keep you in place and to give you further instructions once they are decided upon. You do get to interact with others who have made the trip at around the same time. Our triad are three such prominent figures who fit this bill, all three having met their demise in 1955: Teilhard on April 10 of that year, Einstein on April 18, and Mann on August 12. Their *Kaffeeklatsch*, purposely employing the German for two of the three, ran something like this.

Host: I would like to introduce you to one of our own and one of our greatest. You know it as a muse, not nine as in the mistaken Greek proposal, but who is actually a special envoy of God having the portfolio of Creativity. In its presence we are not to use prose. So tell us about your lives and what you think was left out or was unfinished. Who would like to start us off?

Teilhard: Is dis to be some type of terapy session?

Host: Please Pierre, we are operating in English here, so kindly enunciate your "th's." No volunteers? Well then I will establish the pairings. Thomas, you will comment on Albert, Albert on Pierre, and Pierre on Thomas. I figure we should begin with our literary representative.

Mann: All right, bear with me. Long time reflector, first time commentator.

> May your renown endure in physics for sure
> in science you do take your place
> spacetime and its curve and the laws to conserve
> a paradigm shift do you trace.
>
> Over Newton you stand relativity in hand
> such an imposing figure you cut
> the cosmos was framed it does not look the same
> the speed of light absolute, but
>
> I have a bone to pick and could make it stick
> about the two dreaded atom bomb blasts
> both of these drops killed humans and crops
> and how long the devastation lasts!

> The letter you wrote and what it connotes
> failed to make a modest-sized ripple
> for Presidents I feel were asleep at the wheel
> and could care less about all of those people.
>
> The effort you gave for the humans to save
> I reckon you should have tried harder
> this nuclear fission I hold in derision
> it's tough not to think it was murder.

Host: Thank you, that will do. It got a little personal there at the end. Next.

Einstein: I am not used to this, but if you insist. Sorry, that's not the beginning of the piece.

> A French priest can be in archaeology
> though your Jesuit friends frowned upon it
> to China you were sent as a punishment
> Peking Man rose through your tool kit.
>
> The God in your view is almost totally new
> whom some Catholics could not resolve
> the end for whose joint is the Omega point
> to which God and humans evolve.
>
> From your tradition I quake for it burned at the stake
> those who failed to comply with the rules
> but progress is the game and process its name
> you might have thought your colleagues fools.
>
> With your God I cannot concur, for Spinoza's I prefer
> for whom the world is not subject to dice
> for the thoughts you hold dear you might wish Whitehead were here
> to join forces against your police.
>
> From a mystical me to spirituality
> unlike Whitehead yours a system is not
> while not esoteric you could be heuristic
> so we could all grasp your concepts throughout.

Host: These are getting critical, but that's your choice. This is more of a roast than a toast. Please proceed.

Teilhard: Perhaps I can lend my priestly training to be conciliatory.

> Your rise to fame and widespread acclaim
> mark you a stellar literary author
> the tales that you told as the stories unfold
> in a style compared to no other.
>
> But having said this I would be remiss
> if I did not express my disquiet
> in content and form often outside the norm
> I'm surprised anyone would apply it.
>
> I could not find out what they were all about
> in theme I could find little clue
> from life with a dog to a travel log
> I would rather have my Camus.
>
> Aristocracy was your enemy
> in its ugliness and beauty
> the German life with all of its strife
> I felt like I was watching home movies.
>
> Biographical or no should we consider it so
> could the characters be you or me?
> if my report is too stark and I have missed the mark
> then all I can say is c'est la vie.

Host: You appear to be missing the spirit of the exercise. But this looks like fun. Let me have a crack at it.

> All motion must hang on the old big bang
> to trace it there we trust
> but few elements emerged from this big surge
> to look elsewhere then we must.
>
> From the outward burst, hydrogen came first
> then helium and lithium later
> but this early stuff was not nearly enough
> to construct all we find in matter.
>
> When matter condensed and stars coalesced
> offering an oven-like interior
> heavier elements to make in the large stars did bake
> it took time but was superior.

These stars were hot longer and the procedure stronger
than the big bang by itself could muster
many elements introduced up to iron produced
but not yet gold with all of its luster.

The next stage was set for the rest to be met
as these stars finally went nova
propelled out into space up to uranium took place
the era of chemistry took over.

Physics took aim these elements to claim
and their atomic structure to found
but to take our stance with a major advance
they must to molecules be bound.

Atoms in combination into molecular formation
then in the solar system were stored
when the inorganic yielded the organic
a drop from the Life-Giver was poured.

Many species came alive and many could thrive
and had their day in the sun
but they did not last beyond ages past
and faced their extinction.

Defying the bets amid all the threats
some organisms beat the odds
from simple some moved to complex and so proved
that either could be given the nods.

For those with a niche could be counted so rich
as to increase their longevity
but regardless the sum their time would succumb
to paleontology.

Things became worse in this vast universe
when some perils came from the sky
there was reason to fear as asteroids drew near
and the dinosaurs bid us good-bye.

Multicellular they came *Homo* by name
and eventually stood upright
from the trees they descend and to crops they did tend
leaving nomads in their hindsight.

Sedentary they stayed and caloric they made
their lifestyle to stay in one space
civilization emerged from their population surge
priests and law from one central place.

Think if you will if you've not had your fill
on how mind could arise from non-mind
to posit a core that was not there before
could put us in a real mental bind.

One is prompted to ask whether all of these tasks
with their need of energies redoubled
along with consciousness could be claimed "what a mess"
was it ultimately worth all this trouble?

Now you've got me doing it. I see that Creativity has left by reason of futility. That's not the beginning of another stanza. I imagine we can revert to prose. All three of you mixed praise with criticism, which is understandable, ambiguous creatures that humans are. But this is a place of resolving issues, settling disputes and redressing grievances. Once these are sorted and ironed out, and it might take some time, then we can move on.

Teihard: To what, the beatific vision?

Host: Ah, Pierre, ever the theologian. That depends on how far we get.

Teilhard: So you ask these teleological questions too. Does God think the human experiment was beneficial?

Host: There are secrets hidden from us as well. God continues to delight in humans for reasons unknown to us. Perhaps it has something to do analogously with a sculptor who is not satisfied until the job is seen through to completion and the finished work of art emerges.

Teilhard: I noticed in your verses that you included neither how we became human nor why only one human species, namely us—*Homo sapiens*—survived.

Host: First of all, please realize that this is not our department, but we can shift the question in another direction. Your survival does not mean that you would have been the last and highest development of evolutionary history. One major lasting import of the mythical Genesis 1 tale is that humans came last on the scene, and all which preceded it is that which humans were called upon to manage in a stewardly way. Our colleague Creativity drew humans to exercise some of this custodial spirit which had been bestowed upon and invested in them.

Yet as to your questions, I can offer some more analogies. Recall the life of the deaf, dumb, and blind woman Helen Keller. It was not until strenuous effort was extended to her education by her teacher that it finally got through to her that the name we give to an object represents that object and makes it easier for us to make reference to it once we agree on the designation. It is for good reason that the mythical Genesis 2 account mentions that humans were given the task to name all of the creatures. We are not only story-tellers but name-givers (and regrettably also name-callers). It was the next step for Helen Keller to recognize that these names could also be self-referential and that she in fact was named as such, in essence, "this is me." A similar thing also happens to toddlers between the ages of two and three (excuse the rhyme).

An analogous spark of awakening for this "aha" moment occurred in you at several points ever since there were several human species on the scene. This brings me to your second question. In the creative process, aspects of most other human species worked to make you through their combined contributions. Your genetic complements reveal that you have features, albeit small percentages, of other human species, save those Africans of the sub-Sahara who did not venture out of the continent. Some advantageous characteristics you inherited through interbreeding with them. While true that a new species prevents it from interbreeding with its parental stock, this does not occur immediately.

As you branch off from a common ancestor, in the short term you could still interbreed with them and produce hybrids,[7] though not in the long term. The advantage of this is that at times what you inherited from these other species is what permitted them to survive for the length of time which they did, and you for potentially even longer given the additional advantages you accumulated. In certain instances this came in the form of boosting your immune system. Sometimes they bore the immunology which contained that which yours did not, thereby augmenting it. It is not so much that you and they are hybrids but that you are composites of salient features of your then contemporaries and predecessors before them, all contributing to the genetic pool or pot. In practical terms, God encouraged the meeting of separate human species as a way of piecing together a more robust genome, and you are it.

As far as an analogy is concerned, consider the English language and culture. It is an amalgam, that is, it draws on several sources, of the influences of other peoples. The language exhibits contributions from Greek, Latin, French, Norse, and so on. English is a low Germanic language, and in incorporating features from other tongues and thus becoming a distinct language, this is analogous to our genome where it is not so much that

---

7. Stringer, *Lone*, 34, 197.

others died out in order that you could take their place, it is more like the encounter of multiple persons gave rise or birth to you. You are as you are genetically from the sharing of aspects of others. This is how Creativity has worked, with what was there already so as to make an enhanced product, and you are the result.

You are as you are through the involvement of others. It is not as though Creativity was experimenting at every turn through a process of trial and error. Instead, it was a mixing of traits until the right one came along. All of the others were good in their own way and as far as they went, but they did not need to meet their demise so as to make room for you. All of you were necessary though insufficient ingredients in your own right so as to create a most advantageous work of art. It is for this reason that you are not the only human species present here. For example, *Neanderthals* lasted almost to your present in geological terms and struggled with similar issues which you did, though not always with the same level of language or conceptual sophistication. Just like the Jewish people were to be God's representatives to the other peoples of the world, in an analogous way *Homo sapiens* are letters to other members of the genus *Homo* community. We have ensured that these letters were delivered. The trouble was that you behaved as though you were entitled to the entire planet yourselves without making room for others.

Einstein: That was a very long-winded answer; we are almost sorry we asked. But allow me to ask the age-old question as to why there is evil.

Host: God permits it so that it can be displayed as to who are God's own. There are those of your kind along with some of those within our own ranks who have gone their own way and have rejected our influence in their lives. The fruits of their labor have often become destructive. As the saying goes, "a tree is known by its fruit" (Luke 6:43–45). God delights in humankind while at the same time finds the human condition and the world situation deplorable. And by the way, God is deeply offended at insurance companies' use of the phrase "acts of God."

Teilhard: Why did God take so long to create using the evolutionary process?

Host: God was not in a hurry. It takes time to produce a work of art.

Teilhard: Then why so much extinction along the way?

Host: God is extravagantly interested in life and wants to give life to as many creatures in terms of both disparity (number of body plans) and diversity (number of species) as possible.

Mann: To change topics, is our section commander going to be barking out instructions to us like some kind of drill sergeant?

Host: Why, do you expect to be recalcitrant?

Mann: No, it is just that I left Third Reich Germany for a reason.

Host: You have nothing to fear, for we do not impose or force compliance for any whose hearts are elsewhere. We do not shut the door to them as much as they to us. In either case, the choice is yours and theirs.

Einstein: What is the place of science in your overall scheme?

Host: Science has been bestowed as a gift. It is a marvelous tool, highlighting wonders to behold. But it remains a resource which in some quarters has been turned into a surrogate divinity with those in lab coats serving as priests. You cannot observe everything even with the sophistication of your instrumentation. Apparatus was meant to be a servant; now you have become its servant. You can see only a narrow bandwidth of light which you call visible. But birds can go beyond this and see in ways not possible for you. Bats can both hear and utter sounds well beyond your upper limit of hearing. Dogs have a sense of smell orders of magnitude more sensitive than your own. It is the height of hubris to believe that humans set the standard of available perception, even with their instrumentation.

You seem to have lost a sense, call it a sixth or even intuition, of boundaries. It is like an electric fence (and that is the extent of determinism), which can be avoided for a lifetime, but venture too close and you suffer. What you do is not determined, nor does the historical shape of things conform to some preordained plan, but as in your sports of football and soccer, play stops momentarily once you go beyond the sidelines. You have lost touch with these sidelines. Science cannot help you here. You fail sometimes to sense where they are, or, if you can, you ignore them to your own peril. What is needed is a reignition of this sense.

Mann: I recognize that my colleagues here would be disinclined to ask the following question, since the very issue is bracketed for Albert in that, according to his view, God does not play dice with the world, and for Pierre church authority reflects heavenly authority and would not want in an insubordinate way to usurp it. But I have no such allegiances, so I can feel free to pose it. Do your angelic compatriots ever wager amongst themselves on outcomes, as to whether those humans to whom you are assigned will or will not undertake a certain course of action or will seek to evade it? If God were to predetermine events, then wagering would amount to being unaware of God's plan, or, alternatively, God does not predestine a path for humans, thereby making the wager authentic, for God might not even know the result with certainty and it would not be a legitimate wager otherwise.

Host: The answer is shorter than the question. Wagering is not an issue, since we focus on prompting humans to take the best path for them. We and our associates have our directives and we endeavor to carry them out. God only knows if this is a foregone conclusion; we act as though it were not. As your law enforcement would say, we do not make the law, we just carry it out. We step in typically when the ordinary course of events if left to itself and under its own steam, and of course when we are given clearance to do so, would not produce an adequate outcome. The difficulty to be avoided is for humans to domesticate and treat God alternately as a talisman (good luck charm) and a vending machine.

The trio continued to ply the Host with many and varied questions, some answers for which they regarded as satisfactory while others not. The triad then found that they were out of questions at that point but reserved the right to, what, recall a witness to the stand, I imagine. So once this lull in the proceedings occurred, the three were finally tenderized, shall we say, and in a position to be informed as to why they had been pulled aside and what mission would be laid at their feet—delicate though not impossible—concerning a certain Wolfgang.

Host: Three years after the three of you graduated to this station someone was born who would come to have an affinity with the likes and works of you three. No accounting for some people's tastes, I suppose. But we are not here to make assumptions or evaluations.

Einstein: Being sensitive to matters involving time, why did it seem like the amount of time that elapsed for us to get here seemed like an instant and not three years?

Host: Think in terms with which you are familiar. Similar to a particle or other object travelling at light speed whose clock has stopped and knows of no time interval in the movement from one place to another, you are also not aware of the passage of time when in deep sleep. You are surprised sometimes when morning comes and it feels like you just went to bed. This is how it is like when you graduate to here. No matter how long the time has elapsed between passing on and reawakening, which for the ancients would be on the order of thousands of years if they have not been called here for a mission beforehand, your perception will be as though the lights had just been switched off. In this sense, a thousand years really are like a day (Ps 90:4; 2 Pet 3:8).

The task set before the trio was to be a positive influence on Wolfgang and assist in prompting him in a suitable direction. This did not mean that they were in competition with each other as if to say that whoever convinces Wolfgang to opt for a specific career path most closely aligned with his own wins. No, it is merely that the aptitudes and tendencies which Wolfgang already has displayed could be fanned into flame and nurtured.

Teilhard was up first. Entering a French environment in Quebec as a Frenchman would presumably be least conspicuous, and as having just freshly arrived in Montreal, Wolfgang had not as yet developed a discriminating ear so as to distinguish a Parisian from a Quebecois accent. It was decided that a metro (subway) station which had kiosks serving coffee and sundry comestibles would be a likely spot to assist the two in connecting. And this being an evening time frame, the crowds would be meager and the venue sufficiently secluded. It did not dawn upon Teilhard, however, that the proposal also came with attendant logistics, like was his attire appropriate? Did his timepiece afford the correct time? And did they all come with maker labels and tags (fashion being of crucial significance in Montreal)?

One item of immediate concern was his shoes—they squeaked. Not terribly, but definitely audible and stemming from him. He kept on telling himself it was not noticeable, but when a critical mass of heads turned so as to triangulate the offending sound, he elected to sit at a table. The episode made him think that afterlife standard issue clothing had quality control issues. He pretended to be interested in a newspaper on the table and came to realize that it contained information about a world he had never known and as such became immersed in the photos and articles, so much so that he almost missed his contact opportunity.

True to form, Wolfgang came by and chose to sit by Teilhard's inviting and welcoming face. They chatted about studying religion while keeping one hand in the sciences, extolling the virtues of both, Pierre thereby making his theological-anthropological case that questions of human nature are of utmost importance. They enjoyed an animated discussion which itself presented some problems. Gone are the days when people would look oddly upon others having what appeared to an observer as a one-way conversation, only later to discover that there in fact was an interlocutor on the other end of a mobile phone. What occurred here though was beyond the pale. Onlookers noticed that Wolfgang, unbeknownst to Pierre, was having a discussion with no observable person in particular. It struck Teilhard that he in this state was not widely visible but only to Wolfgang. Pierre evidently did not get the memo. This made passersby believe that Wolfgang's actions could be deemed to render him certifiable.

The two met this way several times until the Host reckoned that they were not getting any further ahead with Wolfgang, plus that their encounters could not remain inconspicuous, as Wolfgang was developing a reputation of being that weird guy in the metro station. Hence things could not go on this way without creating what the local constabulary might call a disturbance. Teilhard found it curious that the heavenly authorities would be so concerned about what the terrestrial authorities might regard as noteworthy. I mean, who runs this show anyway? Pierre then awakened to the thought that heaven, otherwise known as the superstructure, did not aim at autocracy, but definitely attempted to influence the course of events. In its estimation, the superstructure deemed Teilhard's efforts to have been exhausted, though it did not imply that he had outlived his usefulness, nor did the play on words escape them. Seemingly, one did not need to be alive in the conventional sense in order to be effective.

Now it was Einstein's turn. Wolfgang came under the spell of two authors who came to be the subjects of his dissertation. Alfred North Whitehead offered a new understanding of God, but of equal importance to Wolfgang was that both Whitehead and the other main figure—David Bohm—rendered new ways of interpreting the world. Whitehead objected to the notion of the substance view of reality which science supports and chose experience as the exclusive makeup of the universe. Objects in reality do not influence each other by colliding as billiard balls do, rather subjects prehend, or feel, the experiences embedded in other subjects having become objects. In essence, objects are internally instead of externally related to each other, or, in Teilhard's preferred terminology, science observes the world tangentially, whereas subjects develop (think in evolutionary terms) by radiating outwards from their interiority.[8]

Bohm, as a protégé of Einstein, adds to the discussion that reality is multilayered. He maintains that science addresses the explicate or unfolded aspect of existence, while there is also an implicate order where objects are enfolded, and in this state they can be enfolded at one point and unfolded at another even distant from the first and without a time interval in between.[9] (The nagging difficulty with Bohm's orders, sadly, is their undetectability, for no conceivable experiment to date could demonstrate their presence.) This non-locality, as it is termed, prompted Wolfgang to ask the question as to how we can be certain that once, say, an electron disappears from one energy level or shell and instantly reappears at another, seemingly contravening the absolute of the velocity of light, that we are dealing with the same electron? In actuality, contrary to Einstein, there are two sets of laws required to describe what occurs in the micro- versus the macroscopic worlds. Einstein as the last great Newtonian believed that one set of laws could cover the entire cosmos and was thereby suspicious of the conclusions colleagues were drawing from quantum mechanics.

---

8. See previous volume.
9. Bohm, *Wholeness*, 188–236.

Wolfgang was impressed as to how the two sets of laws could be so different from each other, and if they were a married couple seeking a divorce, they would cite incompatibility as the rationale. The microworld laws permitted faster-than-light travel of an electron between shells where the electron traverses the distance between levels without ever being between them. It is never on the way from one level to the other as that would require a time element for it to be part of the way across. This is what occurs at the quantum scale where energy is not infinitely divisible as Sir Isaac Newton suspected, but comes in discrete packets. This discreteness means that an electron can never assume an energy level in between as there are no such levels in nature to assume. Thus the energy levels and the electrons that may occupy them cannot be other than they are, for these are the only ones available. It might not be satisfactory simply to respond to the question as to why this should be the case with "that's just the way things are," but it appears that is what we are left with.

This was a puzzle to Wolfgang. How could a single electron behave this way? For Bohm, his implicate order describes how this can operate. The transfer of a signal at faster-than-light speed telling an electron where to unfold can be accommodated using his strategy. For particle physicists, this is simply the way of the microworld; for Bohm an additional order is called for and its mechanism addresses the problem head-on. So Wolfgang decided to contact the physics department on campus by phone and spoke to a representative, let's call him Professor Edelweiss. When Wolfgang posed the question to him, he was emphatic that it was the same electron. Wolfgang pursued the matter (pun intended) and asked how this could be determined (a quantum witticism). What features of an electron could betray its sameness—does it leave fingerprints? "These are not required" was the only answer given. Wolfgang did not have any emotional investment in the idea that the electrons should be different; he merely wanted an explanation as to why and was disappointed with the peremptoriness he received.

As it happens, this event was also orchestrated by the superstructure. The person on the other end of the line was not any Dr. Edelweiss but Dr. Einstein. Here only Albert's voice was needed without any video component, his trademark frizzy hair would otherwise have been a dead giveaway (pun also intended). This accounts for the brusqueness of Albert's responses, since he exhibited a blind spot concerning the quantum world and could be dismissive about it. He was just being Albert. He could not help thinking that he did not assist things along by allowing his personal feelings to get in the way of his main assignment of prompting Wolfgang. The Host decided that Albert's involvement would henceforth be relegated to an advisory role. Indeed, both Albert's and Pierre's subsequent influence would be confined to the written works of the many and varied popular physics and anthropological authors Wolfgang would read throughout his days. These were not to be understood as demotions as such, just a set of plays sent in by the coach.

Be that as it may, there was a second issue involving Einstein that fired Wolfgang's interest. Time travel, specifically into the past, was something permitted by relativity. The reasoning went like this as Wolfgang understood it. Newton saw the velocity of an object as not having an upper limit. This as we noted was in error as Einstein informed us—the speed of light is an absolute beyond which nothing in the universe can travel. This became precisely the point, for it is true in terms of objects in space but not for space itself; space is under no such restriction. The significance is that faster than light speed is already built into relativity and time travel may very well be reproducible for a particle at the quantum scale. Where we reside, however, seemingly forbids it as an open possibility for us. There appeared to be albeit slight common ground between the relativistic and quantum worlds after all in terms of faster than light travel in both large and small scales, but Einstein still declined the invitation to come aboard and was unwilling to take the quantum view seriously. Wolfgang had met up with stubborn Germans before; his relatives facilitated that.

But there is one more player who could take the field. The final segment involving Thomas Mann, though, was somewhat more multifaceted, and for this we need to go back to Wolfgang's high school days, and it could be traced back even further. In addition to Mann and other German authors, his favorite is and continues to be Indianapolis' own Kurt Vonnegut, Jr. (himself of German descent), whose numerous works were a joy to devour, along with a nod to Britain's own Douglas Adams. At one time he also liked Georgia's own Flannery O'Connor, but later in life decided that he was put off by the brutality of her material. If you want to know about the dark side of human nature in need of, let us say, spiritual infusion, Wolfgang was known to put forward, then he recommends reading Salinas Valley's own John Steinbeck, whose offerings, in comparison to O'Connor's, in his view, are more compelling and did not leave him feeling cold and parched.

Whereas the exposure to Teilhard's influence was through a form of his specter and Einstein's was through his voice at the other end of a telephone line, Mann's was different still. Wolfgang had grown particularly fond of a used and antiquarian book shop located just off-campus, where Normand the proprietor and he would discuss a wide range of topics, including but not limited to literature but also science, each theme having a portion devoted to it in Mann's works. And when he reflected on it, Normand did have a resemblance to Mann, though with a full beard. Wolfgang discovered that some of the answers to questions of which he was in pursuit also previously animated Mann's thinking, particularly in his *Magic Mountain*, such as the nature of life and the movement from inorganic to organic material, together with inquiries into the concept of time. Along with these topics also arose religion and its place in society, and even psychical research.

As for the latter, this occurred during the time when Wolfgang hit a peak of interest and involvement in it as a field of academic study. Placed in the broader category of the relationship between mind and matter, it yields adventurous excursions into what lies beneath the surface of ordinary human abilities where the extraordinary can be tapped. Few of us, it was concluded by the two interlocutors, come with full-blown deployment of perception in extra-sensory terms, but few of us again are entirely without some experience of it. Mothers, for one, appear able to intuit their own child's cries amongst a welter of them and when the child is in distress. Psyches perhaps could be mined for these resources. We err if we assert that any such access is denied to us or that God's Spirit never avails itself of these means at its disposal. The difficulty surfaces in discerning when baser elements in the so-called black arts are counterfeiting them. This became an issue Wolfgang never resolved.

Nevertheless, literature was not the only art form that had an impact on Wolfgang's outlook. Music, as it does for many persons, also became inspirational. The pleasure of listening to a performance of it was a feeling both he and Mann, at least in the form of some of his fictional characters in *The Magic Mountain* and especially in *Doctor Faustus*, shared—being enraptured by the kind of music that touches the soul. Classical music in particular is a major theme in the latter, as Mann has his characters greatly moved by it. Wolfgang's initial foray into the world of music occurred as a youth many moons ago (Indigenous metric) when he purchased The Beatles' "Hey Jude," Credence Clearwater Revival's "Up Around the Bend," and Jimi Hendrix's "All Along the Watchtower," written by Bob Dylan. What inspired him was both the artistic ability of the musicians and the strength of their lyrics. Sometimes the latter had a political edge to them, as in The Who's "Won't Get Fooled Again," the tune which was to become Wolfgang's lifelong favorite.

At many a turn, the feeling of raw power from electric guitars in particular, especially when the volume was "turned up to eleven,"[10] seemed to release the brain's own opiates. Thankfully the experience was not addictive; he just sought it out in any spare moment, located in his own personal sound room to which he could retreat whenever the urge hit. What it also afforded him was a safe pastime (in addition to sports, aside from injuries) so as to keep him out of trouble.

---

10. A classic line from the 1984 film *This is Spinal Tap*.

Music continued to be a source of enjoyment and inspiration, but it was not until Montreal where it began to manifest in his writing skills. He entered his program with low writing caliber, but as the program progressed the quality increased. Creativity appeared in ways not previously experienced. Montreal, as it turns out, at least at the time, played host to the world's largest annual jazz festival, which Wolfgang and his wife attended for twenty-one consecutive years, well beyond their Montreal period. He was buoyed by the euphoria, almost narcotic in dimension. Now this is not to suggest that an annual injection of inspiration of this type was effective for the duration of one year; it was just a major contributing factor augmented by nearly daily doses of music at home.

It could hardly be denied that artistic experience is contagious and leads to inspiration in one area or another. This has gratefully continued for Wolfgang ever since. Again, this is not to suggest that literary ability would not have surfaced without music, or that a steady diet of it is a necessary ingredient without which a literary tendency would not flourish; it simply helps. The prevalence of artistic capability yields more of the same or similar for some persons. By the time Wolfgang departed from the Montreal chapter of his life, he had developed a style of expression that emerged even without his notice. He only became aware of it once his supervisor mentioned it to him. It would be hyperbolic to claim that Mann had a direct influence here, but that he had an influence is undoubted, along with that of many other authors who helped fan the spark into flame. Wolfgang was once counseled that if you want to write well, read capable authors and write as much as you can, even if they are just letters. This he did, and it worked. And so did music. Feel free to attempt this in your own domicile.

Now that each member of the trio has had an opportunity to extol the virtues of his craft to Wolfgang, they can safely send him on his way having completed their mission. It was not so much that Wolfgang was channeling the triad as each of the three was a conduit for Creativity, all pointing in Wolfgang's direction and making for a triple-barreled influence.

This has been a work of fiction together with non-fiction. Which portion fits into which category is clear only to Wolfgang, and he is not letting on. In reporting on this research, I could have started off with "There once was a man from Canada," but that sounds too much like the beginning of a limerick. Besides, not much rhymes with "Canada" ("America?" "armada?" "rotunda?" "gondola?" "cinema?" "enema?"). How seriously do I take this investigation into one person's experiences? Only in unguarded moments, but I will not divulge how often I encounter them. And as for Wolfgang's identity, all I can respond with is, need you ask?

PART 1

Religion

THE BIBLE: EDITED VERSION

One interesting feature of the ancient biblical languages, mostly Hebrew, Aramaic and koine (common) Greek, is that they had an almost complete lack of punctuation. Not that they had a blatant disregard for it, but that these languages did not afford many. Readers who first stumble upon the documents are taken aback, not knowing where and when sentences and paragraphs begin and end. Furthermore, the punctuation marks which may have haunted us during grammar class, such as periods, commas, colons, semicolons, question and exclamations marks (and don't forget brackets and parentheses, much like I am using here), we later came to appreciate for seeing to it that our meaning comes across, with the avoidance of lower grades at best and international incidents at worst. Those who translated the above texts into Germanic languages, which includes English as low German, assigned sentences and paragraphs in places where they perceived the authors had completed their thoughts. Chapter and verse designations were also applied.

At times, ancient authors might insert a parenthetical note which they hoped would produce a proper segue that the text as it stood may not, or inject a tidbit of information in case cultures approaching the text were unaware of traditions or observances of foreign nations. Later, when modern translators introduced punctuation into their versions, some passages were set off in parentheses, believing these were asides or off-hand remarks geared either to enlighten or bring back to recollection certain beliefs and/or practices of yesteryear. Another instance of editing by way of parenthetical information was by theologians who intended to campaign for their pet doctrines which they thought were inadequately presented in the text, and they sought to ensure that readers would not fail to make the connection.

Admittedly there are times when the biblical authors are given to editorializing, such as in 2 Kings 17 when Israel was taken into exile by the Assyrians. The author(s) pivot from history to theology when they present the alleged reason as to why all of this occurred, namely the wrath of God upon Israel's idolatry. But most asides do not operate in an imposing way. Instead, there is little reason to doubt most of them or treat them as unreliable. That the season of spring, for example, is the time of year in which "kings go off to war" is not the kind of fun fact that appears to be agenda-driven, such that were it to be omitted, the message would be imperiled; it merely serves to flesh out the narrative and offer some historical background. Notably, that statement, while parenthetical, is not set apart by parentheses. Nevertheless, it may be the case that asides are among the most reliable aspects of an otherwise partisan document. Either way, the original author had a message to convey as did later redactors (editors), and sometimes these two were at odds, with the latter seeking to "correct" that which would otherwise likely be misconstrued or escape the attention of the reader. Accordingly, an alternative title for this segment could be "These Statements Aside."

Before we begin, please understand, as always, that we undertake all such exercises with the scriptures as they stand—*prima facie*—at first glance, for purposes of argument. We permit them to have their say for the moment. Plus, should there be Christians who fail to entrust the exegesis and interpretation of their scriptures to non-Christians, how would they respond to those Jewish scholars who feel the same toward Christians vis-à-vis their own Hebrew canon? Would these Jewish folk be justified in demanding emphatically, "Hands off our, what you call, OT?" The Christians might reply by claiming those writings as having been inherited by them as the true Israel. Convinced? Can they really have it both ways?

Back to the issue at hand, after an initial flourish into the possibility of a computer search engine enabling me to inquire into the Bible by way of punctuation marks, in this case parentheses, and being rebuffed for my efforts (they can send a person to the moon, but . . .) I set about the tedious task of consulting the scriptures manually, column by numerous column, hoping my eye would fall upon and catch the wily marks. I then compiled a list of as many as I could locate, followed by slotting the items into categories according to context. My comments will be directed to the type and amount found in each theme. I conducted this investigation in order to determine which content could be considered agenda-driven and which off-the-cuff remarks or statements of fact.

Sometimes it was not entirely clear as to which categories the passages should be inserted into, and some could be slotted into multiple. There are also many more asides not surrounded by parentheses, but the examination here will mostly be confined to those which are. The categories total fifteen in number. I shall commence with those categories containing the least amount of passages and proceed toward those with the greatest. It should be noted that the NRSV uses roughly 39 percent of those parentheses found in the NIV, which themselves total approximately 225 by my reckoning, a full listing of which is placed in the appendix. The former is more inclined to use dashes or commas in their place.

Initially, there were two categories of three items each. In the order of their appearance in the Judeo-Christian canon, the first concerns the furnishings of the Temple, all from the OT. One verse is a restatement of another, and the third includes the origin of the materials employed in the architecture as well as musical instruments, namely from Hiram, king of Tyre. This third passage occurs in 1 Kgs 10:11–12 (note that restatements of much information in 1 and 2 Kings occur in 1 and 2 Chronicles, in this case 2 Chr 9:10–11) and appears out of place in a pericope (passage on a specific theme) devoted to the Queen of Sheba's visit with King Solomon. It should have been placed at the end of the previous chapter, after verse twenty-eight, omitting the first point by reason of repetition. The second category covers weights, measures, and currency values, all again from the OT, stating very simply how many of one converts into another, in case someone is not familiar with their usage. One wonders why the author would make a point of informing readers or listeners about a designation that would be common knowledge, and should it have been a practice from history which has fallen out of use, then why was the past tense not employed?

Four items are found having to do with biological information, all once again from the OT. One deals with the color of goats, a second discusses when certain grains ripen, a third mentions when it was the season for grapes, and the fourth, not in parentheses, is Isa 28:24–29, which refers to agricultural techniques and how to thresh certain crops. This fourth is the only one of the four bearing an additional editorial component, stating that this wisdom comes from God.

Two categories contain five and six elements each, respectively: one dealing with customs and the other with divine activity. For the former, one treats the type of jewellery some nationalities wear, the second with previous policies about deed transfer (Ruth 4:7), the third with Jewish rituals for washing and cleansing, implying that his readers were not familiar with them, the fourth states flatly that "(Jews do not associate with Samaritans)" (John 4:9b), and the fifth as to how Athenians occupy themselves. As for the latter category, the first tells of God working in the attitudes of the Egyptians so as to make them more positively inclined or receptive toward Moses and the Israelites, the second with an appearance of the angel of the Lord, the third as to how God was working in the life of Samson to subdue the ruling Philistines, the fourth, on the contrary, how God was working through foreign nations to subdue Judah, the fifth on God speaking through God's prophets, and the sixth (Neh 13:2) on God turning Balaam's oracles (as told in Numbers chapters 23 and 24) from a "curse into a blessing."

The first of the categories containing double digit entries, in this case ten, treats personal data, gifts and skills. The first that the Israelites amount to "(Jacob and his descendants)" (Gen 46:8), the second reminds the reader about the untimely death of Er and Onan, the third about circumcision, the fourth about two who were bestowed talents in the construction of the Tabernacle, the fifth extoling the virtues of Moses, who "(. . . was a very humble man, more humble than anyone else on the face of the earth)" (Num 12:3), hubristic, one would think, if Moses is the alleged author of the work, the sixth and seventh about who owned which tract of land, the eighth and ninth about who married Solomon's daughters, and the tenth about which nation was "(famous as archers)" (Isa 66:19).

Another category with ten concerns translations, this time all from the NT, perhaps because, when written, the Jewish authors of the books of the OT did not expect other nations to take an interest in them; they were written by themselves for themselves. One subcategory involves a command, two: landmarks, three: names, and four: titles, specifically Rabbi, meaning Teacher, and Messiah, meaning the Anointed or Christ. Nine of them proceed in the direction of Aramaic to koine Greek, and only one from the Greek to Aramaic. This suggests that the author(s) intended to educate their Greek audiences in order for them not to miss the significance.

Next in line, with twelve entries, covers negative character qualities and misdeeds. The first with the death of two of Aaron's sons, the second about Abimelech's killing of his seventy brothers, the third about the misbehavior of Adonijah, the fourth about how one good murder deserves another, the fifth about a lie uttered by a prophet, the sixth about the reprehensible actions of Ahab and his wife Jezebel, the seventh about the contemptible practices of an adulteress, the eighth that Jonah was running away from God, the ninth about why Barabbas was incarcerated, the tenth that Judas was a traitor, the eleventh as to how Judas met his end, and the twelfth regarding our need for discipline.

The next category covers lines of descent and contains seventeen items, all from the OT. One reveals the name of someone's father, another focuses on descendants, still another on who the priest was at the time of Solomon and seven each give attention to ancestors and offspring. Important to keep in mind here is that the latter are not necessarily immediately previous or next generation figures, but might very well be relatives more distantly related in time. Geographical elaboration marks the next category, also containing seventeen entries and all again from the OT. Two deal with inheritances and land allotment, and four each with borders, other (foreign) nations, and nearby landmarks. My two personal favorites tell of the high caliber products of a certain country: "(The gold of that land is good; [pearls] and onyx are also there.)" (Gen 2:12), nice to know where to place one's ancient investments; and a distance given in terms of time requirement: "(It takes eleven days to go from Horeb to Kadesh Barnea by the Mount Seir road.)" (Deut 1:2), helpful information permitting the reader or listener to know what to pack.

Skipping over a category for the moment so as to treat it at the end, the following category contains nineteen items and is simply reserved for miscellaneous information. The first states that "(I know you have much livestock)" as the land of Canaan is being apportioned, the second is about a conspiracy struck between David and Jonathan, the third about a book not contained in the biblical canon and which is also lost to history, the fourth concerning Bathsheba's purification from uncleanness, the fifth about the furtive absconding of the bones of Saul and Jonathan, the sixth on apocalyptic beasts, the seventh about the attendance at Jesus' first feeding miracle according to roughly how many men were present, the eighth on Jesus lamenting about the lack of honor given to him and others in his prophetic tradition in their own "hood", the ninth on the name of the servant whose ear Peter had cut off with a sword, the tenth through twelfth about the resurrected Jesus appearing to his disciples by the Sea of Galilee, the thirteenth on the name of the field Judas died in, also constituting a translation, the fourteenth about who was gathered at Pentecost, the fifteenth on an assumption made at the time of Paul's arrest, the sixteenth about Paul's brain cramp regarding who he baptized, the seventeenth on Paul's self-assessment concerning his mental well-being at the moment, the eighteenth about Paul's instructions pertaining to the fellow-worker Mark, and the nineteenth about not imposing unnecessary burdens.

The following category concerns name variations or aliases, twenty examples being found here. One has Moses changing the name of Hoshea to Joshua, two deal with why and that Esau is also known as Edom, one has a son who was not born first yet has been appointed firstborn rights, two deal with prophets and seers, one has Preparation Day as another name for the day before the Sabbath, two have another term for a palace (the Praetorium) and the Sea of Galilee (the Sea of Tiberias), three have Jesus renaming disciples, and eight deal with additional aliases.

The next category, having thirty-two entries, contains varieties of national, city and town, and site names. Summarily, three are for nations, twenty-one for cities and towns, a further two for inheritances, and six for assorted sites including mountains—a veritable salmagundi—several of which make reference to the Dead Sea. The penultimate category deals with historical backgrounds, and contains a whopping fifty-three items, all but the last from the OT. There is one each in six sub-categories: former land names, land holdings, offspring, historical records, the copy of a letter, going into exile, and marriage to foreign women. Four sub-categories have two each: other gods, prophets and seers, priests, and the ark of the covenant. An additional three have three each: divine activity, forefathers (both Jewish and other nations), and the return from captivity. Six concern other peoples, ten with elaborations on kings and war, thirteen about cities and towns, some of which were cities of refuge, and lastly an explanation of the meaning and significance of the original observance of Purim. My two personal favorites in this largest batch are these: "(Formerly in Israel, . . . the prophet of today used to be called a seer.)" (1 Sam 9:9), and as intimated above, "In the spring, at the time when kings go off to war . . ." (2 Sam 11:1).

The vast majority of the foregoing are informative but innocuous, save, of course, for those dedicated to divine activity. The final category contains nineteen elements, meaning less than ten percent of the mostly parenthetical material are not of the factual variety, apart from the eight dealing, as stated, with divine activity. So let's have a look at those items which are more than merely asides.

The lone OT entry makes the editorial point that foreign gods are false gods. Next, a redactor to Mark's gospel turns Jesus' statement that foods do not make one unclean into the personal predilection that therefore no foods are unclean by Jesus' say so, which does not logically follow. There can still be unclean foods, but their effect need not sully the consumer. This later editor could have made a gloss on this passage and inserted his own intent (no need to write "her," since none was encouraged, invited, or permitted to contribute). Not only was this not the intent of the pericope, but it promotes the type of doctrine which that editor espouses and wishes the reader in no uncertain terms to arrive at the same conclusion. Perhaps it was directed toward purveyors and/or consumers of foods not considered kosher, assuring them that they can also belong to the tradition, regardless of the official Jewish stance.

A second item from Mark makes the critical comment on behalf of Peter that he was so taken aback by Jesus' transfiguration that this prompted Peter to run off at the mouth. Yet this is an attempt to psychologize. Following these examples, in John's gospel the emphasis on the part of the gospel author(s) is to hold to the resurrection as of utmost importance. The disciples ran to the tomb and found it empty, and the parenthetical statement is made that they failed to grasp from the scriptures that Jesus must be raised again to life. By way of commentary, the only access they had to the scriptures, being themselves illiterate, was to Jesus' own exposure and interpretation of them. Were Jesus himself to have been illiterate, then the disciples would have needed to depend on his having heard the sacred text had they not heard their delivery themselves. The point to be made here is that it is editorializing to insist that the Hebrew canon, which was the only scriptures available to them at the time, definitively made this particular resurrection claim, and that is not conclusive, outside of the Jewish resurrection alluded to in Isaiah 26:19 and the general resurrection intimated in Daniel 12:1-3.

Paul in Romans then maintains that the Gentiles have an internal awareness of the law even without having been introduced to it externally. This plainly overreaches what one can claim about the inclinations of someone's heart and amounts to a preference on the part of Paul (spot the alliteration). Why, for instance, could they not be driven to embrace a different moral code? Paul takes further liberties with the OT, specifically Deuteronomy, when twice Paul assumes that the passages refer to Christ, which is not at all apparent, and in the second of the two he mistranslates the text (Rom 10:6-7).

On the topic of marriage, Paul offers both a command from the Lord, but provides no reference for it, suggesting that it must originate from some other written source or Jesus' unrecorded sayings, and then admits that he is further submitting his own command which ought also to be given a hearing, much the same as Jesus' own (1 Cor 7:10-12). In the same epistle (letter), Paul as a Jew declares that he is "not under the law," though he is "not free from God's law" while at the same time is "under Christ's law" (9:20-21). What he likely means is that the Mosaic law has been fulfilled in Christ, but what is unclear is how God's law relates to Christ's law—are there two sets of laws, one still in effect though superseded by Christ? These and other points allow Paul to render his theological views of the place of law and grace/gospel in the lives of Jesus' followers, and as such they are deeply held personal interpretations of Jesus' intent as well as those of the disciples with whom Paul interacted.

Lastly, John of Patmos in Revelation gives the meaning of a symbol and the order in which people will be resurrected. The upshot of all this is that one can have confidence in the reliability of parenthetical material, except for the ones ostensibly making interpretations, since they are offered matter-of-factly and as such are not appreciably agenda-driven. Thus if one is searching for biblical statements containing the least amount of propaganda, these are among them. No one's reputation seems largely to be on the line with them, and they appear to supply accepted information, which one is not required to hold in order to be a devotee of the tradition. What is not at stake with them are tenets of faith. And, as mentioned, some items are for the benefit of those for whom certain concepts are unfamiliar, the target audiences often being non-Jewish. This then is the version of the Bible that just happened to be the one which was copied, survived, and transmitted to greet subsequent generations.

WHAT'S THE USE?

Moving from one of the longest segments in this investigation to one of the shortest, let's examine internal biblical concerns together with some early Christian history.

We are cautioned in Eccl 7:16 not to be over-righteous, because the consequences could be such that we could very well destroy ourselves. Was Jesus over-righteous then? What he said and did destroyed him. He died before his time, at least from the perspective of the longevity of an average human lifetime in that age, which was not lengthy, not from what God had apparently planned, according to tradition, for the Messiah to endure. He was not an Ubermensch—an overman or superman—in the Nietzschean sense, a warrior-like figure having the will to power, which is precisely the very person the Jews were anticipating the Messiah to look like, though the Nietzsche reference would have been lost on them, since anachronistic. And when Jesus failed to match this portrayal, the Jews were left asking God, "What else you got?" A crucified Messiah, after all, is a tough sell in Jerusalem, while a resurrected body as good is a tough sell in Athens, for it is precisely the body that constitutes the problem for the ancient Greek world.

Paul, being well-versed in the scriptures, studying as he had according to tradition under the teacher Gamaliel (Luke has him saying as much in Acts 22:3) and was thereby also well-credentialed, was delivering a speech to the Athenians according to the latter half of Acts 17, and when he mentioned the resurrection of the body he lost a sizeable portion of his audience. Some sneered and left, others wanted to know more; what we do not know is if the latter eventually became persuaded by Paul's message. Since neither the Jerusalem nor Athens strategy amounted to much, Paul and the disciples might have been better off consulting a marketing specialist for a different tactic. If you want to attract a following, these consultants may have advised, then you will need to offer a message that will resonate with them. As it turns out, the Jerusalem church did not have staying power. It's a wonder that the Christian movement ever got off the ground.

The story was different in other parts of the Mediterranean. Paul planted many churches which took root and thrived, though the flourishing was not by leaps and bounds, nor was it without persecution in spots. The church grew, against all odds, albeit certain communities were forced to go underground, quite literally. Some detractors accused Christians of drinking blood and killing infants by drowning—a deliberate misinterpretation of the sacraments of communion and baptism. And they say even bad press is good for publicity. Not here.

How could devotees put so much effort into a vilified religious movement? Christians might be heard to intone that the fragile and tenuous survival of the church was the work of none other than the Holy Spirit. If this were to have been the case, then why did the Spirit not make great gains in Jerusalem? Is the same diagnosis right in one instance but not in the other? One figures that God would have had greater emotional investment in what was the Promised Land after all, and especially Jerusalem—the place where God chose to set God's name (1 Kgs 11:36, along with numerous other passages). Or even after their defeat by the Romans in the Jewish Wars, did the God of the Jews still nurse a grievance against the stains in the land perpetrated and perpetuated by these same Jews? And most significantly, let's determine this theologically once and for all from a doctrinal perspective, did the crucifixion of Jesus just outside Jerusalem add fuel to God's fire or quench it?

But speaking of after alls, in the end when there will be a new heaven and earth (Rev 21), the city's name that will be coming down from heaven to earth will not be the New Athens but none other than the New Jerusalem. Hence God has not given up on it just yet, even in the face of some historical appearances to the contrary.

WHO'S ON FIRST?

We turn now to Eccl 3:1–8, and you can probably surmise at this point that, in my appraisal, this biblical book commands a generous amount of wisdom on which to ruminate, and that life affords the opportunity for any undertaking; we simply require wisdom in order to recognize when those times arise. We are informed that the Teacher in Ecclesiastes submits a list of opposites, both of which in each case, and one would assume also the positions in between the extremes, are appropriate given the correct circumstances. Curiously, the list includes times when gathering stones are called for, and others when scattering them is warranted. One wonders what the benefit could be in participating in either act: merely to be prepared like a good boy scout and earn a merit badge for preparation just in case one finds oneself off with the cronies to a stoning, or in the entrepreneurial spirit taking some along and setting up shop with a booth as a vendor on the way to that very same event and offering choice ones for sale to those too feckless to bend down and pick up a handful themselves from the abundance of them in the region?[1] But much more alarmingly, the list also invites contributions from those engaged in killing, warring, and hating. One would have thought that exemplars, at least, would be above those sorts of things. Or are they accommodations for the bellicose among us?

The Bible employs the literary device known as merismus, where opposites extremes are combined so as to describe and stand for a totality, most noticeably here in Eccl 3 and in Isa 45:7, where God announces that "I form the light and create darkness, I bring prosperity and create disaster," intending to imply that God is minister of the entirety—the extremes as well as all points in between. Something similar can be applied to our contemporary context. Governments, apart from the unscrupulous ones, have been commissioned with not only serving their constituencies, but ensuring that order is maintained. This can be accomplished, so runs the argument, in several ways, each involving some form of law enforcement. A spectrum can be drawn for the three main options of legal moralism at one extreme, which lobbies for more laws against immorality to establish greater safety for its citizenry; libertarianism at the opposite extreme, where governments

---

1. A nod to the 1979 Monty Python film *Life of Brian*.

should not limit individual liberties and thus which campaigns for fewer laws, only those securing persons from being harmed by acts on the part of others; and paternalism somewhere in the middle, where the mindset is that some people need to be protected from themselves, or saved from harming themselves,[2] the government acting in a parental or fatherly (from the Latin *pater*) fashion.

The interpretation could be made that all three can stem from a conservative mentality and these folks may be found at any of the three points on the axis: those who seek to be tough on crime in the first position, those who would pursue the ideal of smaller government in the opposing terminus, and those given to a father-knows-best presumption as the mediating stance. Some governments of varying complexions have operated in these manners, sometimes even within the same administration. And for those governing bodies whose intent has been expansionist, colonialist or even imperialist, wishing to export their culture to other lands and peoples not asking for them, from a because-they-could-really-use-it mentality, would endorse the middle camp where they could be heard chiming, "we know and have the right and better way for you."

Both governments and their people have taken on each of the three of these methods when dealing with the Covid pandemic, even fluctuating between them. At times, ruling bodies have decided that they require more extensive lockdown measures so as to curtail the spread of the disease. In such cases, this has come with stricter laws and harsher penalties, reflecting the legal moralist alternative. In response, some groups have protested that their rights have been infringed upon and have sought an easing of the restrictions and protocols so as to resume greater freedom, in line with the libertarian approach. When these efforts have failed to flatten the curve, governmental rationale has sought to be parental toward the people and reinstituted some policies "for their own good," as the paternalist would invite.

Each solution has had its successes and failures. For those nations having experienced the former, their methodology is questionably understood as automatically transferable to other contexts, though certain priorities might not translate precisely into an optimum path when applying what other nations recommend. Following their example might not readily be patched into the tear in the fabric of other countries. We may ourselves desire that the entire planet would have adopted a lockdown regimen from the outset in order to eliminate the threat before it ever proliferated. Others would rather have taken a combination approach, mixing an evolutionary strategy—"sometimes we need a culling of the herd to determine who is fit

---

2. Holmes, *Ethics*, 105–6.

for survival, thereby leaving the remainder behind" (a truly uncompassionate position)—with a theistic one—"and God will take care of the rest" (on its own a somewhat naïve viewpoint).

What can be evaluated about how each population has handled the virus and its variants is how well it has displayed consideration for others. Those with a me-first individualist perspective might probably have battled the pandemic for prolonged periods (the U.S. for example). Those who have temporarily set aside personal interests for the sake of others may perhaps have come out on the other side (New Zealand for instance). Here we observe the distinction between conservative and liberal: the former are content to compete with the disease so long as they can live their lives as normally as possible; the latter are willing to put up with the inconvenient and abnormal steps taken in order to regain their foothold.

I wonder if the Teacher in Ecclesiastes would have concurred that there is a time (and a place) for each of these alternatives. Would he have approved of all the measures adopted thus far? Would he have given his approbation for emphasizing at times either self or others? Maybe so, when it comes to individual front-line workers sometimes needing to consider their own as well as their family's welfare. But as a nation? Which begs the question, was the Teacher writing to individuals or collectives or both?

And while we are on the topic of Ecclesiastes 3, and as an amplification from last time, verses 18–21 inform us about the following five notions concerning human nature: (1) we have continuity with the animals; (2) they and we have the same origin and can expect "the same fate," namely the dust; (3) because they and we "have the same breath;" (4) dualistically, they and we have a spirit, which is alien/foreign and not native to each; and (5) together with 12:7, the human spirit, at least, "returns to God who gave it."

ADDENDA

There is a Miss Conception (a Spanish woman) that Miss Cellaneous (either a Portuguese or Italian woman) material is unimportant. I am here to dismantle this erroneous idea with a few tidbits of my own, beginning with theological items and ending with philosophical. Initially, there is a tradition within Christianity that insists that grace in general and salvation in particular is not something that can be earned or merited; it is simply a gift. This is a quaint sentiment but it does not cohere with the thrust of biblical evidence. For better or for worse, we are informed as to acts we are to perform and obligations we are to meet and fulfill. John the Baptizer, according to the text, instructs his hearers that they should "produce fruit in keeping with repentance" (Matt 3:8; Luke 3:8). When they asked what specifically it is they should do, John supplies practical counsel: to the tax collectors, for instance, John tells them not to exact more than they ought, and to the soldiers he implores them not to extort or give false testimony and to "be content with your pay" (Luke 3:10–14). Notice both that these exhortations are social in general and often economic in particular, and that he never chides the military contingent to refrain from being soldiers, implying, against a pacifist view, that it constitutes a legitimate vocation.

The letter of James adds to the discussion by urging that "faith without deeds is dead" (2:26). While works might not be entirely salvific, we are "justified by what [we do] and not by faith alone" (2:24). There are those who will claim that John's teaching as in the above is still part of the old covenant, since Jesus had not as yet begun his ministry, and as such must be understood as superseded by the new. Well, James *is* part of the new and as such compels attention and action on the part of God's followers. Flatly, the tradition in view is at least partly in error, human product that it is. Its heart may be in the right place, shifting as it does the focus from our abilities to God's accomplishments, but its head has demoted a portion of the scriptural witness, the totality of which it otherwise upholds.

Secondly, Job 1:6 and 2:1 mention that there is a council in heaven acting as servants of and possibly even advisors to God and is populated by "heavenly beings" (NRSV) or angels (NIV), also known as (a.k.a.) "sons of God." This highlights the fact that the appellation "son of God" in relation to

Jesus was not the original instance this occurred in the scriptures (but was actually (chronologically) introduced when God referred to Israel as being God's son (Hos 11:1) and (textually) when God informed David that his son Solomon would also be God's son (1 Chr 27:6)). This may also have ignited Paul's at least erstwhile implication that Jesus' previous existence might very well have been angelic: "you welcomed me as if I were an angel of God, as if I were Christ Jesus himself" (Gal 4:14), or so it could be interpreted, and perhaps Paul did. It depends on whether the second clause in the Greek amplifies the first or is a restatement using an in-other-words approach as it would be in the Hebrew.

Thirdly, as a warm-up exercise, and to refresh our memories that discrepancies and inconsistencies do in fact occur in the sacred text, there are fifty-five Psalms attributed to David up to but not including Psalm 72, which instead is from his son Solomon, where in the last verse it is stated that "the prayers of David" terminate (verse 20). It seems odd, therefore, that from Psalm 73 to 150, the final one in the canonical text, there are seventeen more attributed to David, which becomes a nuance of "terminate" that must be adjudged as alarmingly gratuitous. But this is not the thrust of our intent here. What fires my interest is the character qualities Jesus bore, what kind of dude he was, and if we would have wanted to hang with him. Given the number of banquets he was invited to and attended ("the son of man came eating and drinking . . ." (Matt 11:19; Luke 7:34)), then he did a fair bit of hanging. Banqueting might have left him a touch on the pudgy side, but with the amount of walking he and his disciples engaged in, spreading his message throughout Jewish, and one Gentile, territory, his Fitbit would have worked overtime. Perhaps this was the cause of his healthy appetite. The impression we get from the accounts, in any case, is that if you were inviting Jesus to a meal, you had better hurry, since his dining schedule was quickly filling up.

We are given glimpses as to what Jesus was like and he comes across as an unflinching hard-liner when he informed his hearers that their quality of life needs to surpass that of the ruling religious authorities if they seek to be admitted into the kingdom (Matt 5:20); even calling someone a fool can endanger them of severe judgment (verse 22); and similar statements occur on the issues of adultery (verses 27–30), divorce (31–32), oath-taking (33–37), retaliation (38–42), loving one's enemies (43–48), almsgiving (6:1–4), prayer (verses 5–8), fasting (16–18), where our true treasure lies (19–21), judging (7:1–5), the road to life (verses 13–14), self-deception (21–23), and many others. He is showing us a better way, which in itself is a positive thing, but he does so in a threatening way—do this or else—as it were. He brooks no perfunctory followers but only the dedicated. One needs to be committed to his cause; all others need not apply and whom he refers to as "the dead" (8:22).

This is because discipleship is difficult (10:26-33; 37-39), it tears up households (verses 34-36), even resulting in distinctions between one city and another (11:20-24), and after all this he incongruously proclaims that he is the source of rest (verses 28-30). Yet on the heels of Peter declaring correctly as to Jesus' true identity, no doubt feeling pretty good about himself, having just become the winning contestant on "Name That Messiah," he imagined he was in an accomplished position so as to deliberate along with Jesus on his mission strategy, since he reckoned he was in the innermost part of the inner circle and that Jesus had lifted him into an advisory role, announced that Jesus' fate would not be as he described it. Peter was then met not with "That's a nice sentiment and you are a dear to mention it," but with Jesus turning on Peter and detonating with the exclamation, "Get behind me, Satan!" (Matt 16:13-23; Mark 8: 27-33) Satan? Really? Jesus apparently thought that Peter was perilously close to unwittingly derailing Jesus' mission, like the devil would attempt to perpetrate.

This makes Jesus not only assertive but adversarial; best not to get on his bad side. Nowadays people might say that he needs to chill out, though that is not the way he saw it. At times he could be an eating and drinking buddy; at others he was focused with his sights trained on the task set before him. There are instances when Jesus did exhibit a gentler side as when he welcomed children to him (Matt 19:14; Mark 10:14; Luke 18:16), and entered Jerusalem on a donkey (Matt 21:7; Mark 11:7; Luke 19:35). But this is about the extent of his meek and mild portrait. On the other occasions we are informed about his teaching and activity, and they are not exactly whimsical. This could be owing to what the gospel writers elected to report on, and given that Jesus did not have a ministry protracted in time, stretching from between only one to three and a half years, perhaps he needed to give strenuous attention to the urgency of his message.

Fourthly, certain constituencies within Christianity promote family values as something which is a biblical norm. The most conservative among them will also declare that a nuclear family is composed of a husband, a wife, and kid(s), thereby intimating that same gender households are an aberration and are not to be considered as enjoying divine approval. One of the troubles about this is that one should not appeal to Jesus as promulgating family values. There are times when he would not give his mother and sibs the time of day as when they were asking for him, only to have Jesus declare that the crowd presently listening to him and preventing his family from reaching him actually constituted his family (Matt 12:46-50; Mark 3:31-35; Luke 8:19-21). Not only that, but Jesus saw it as among his tasks to bring division as opposed to peace between family members. In place of family harmony, "the offense of the cross" (Gal 5:11) is geared toward turning the members of a family against each other (Matt 10:34–36; Luke 12:51–53). Family, contrary to the perspective of some, is not the most important thing for Jesus. Blood relation is not the main element for him, as it can be in the eyes of some Jews who trust in their father Abraham (and in another sense the Temple), but in following his example and teachings.

Fifthly, on the topic of Jesus and the Temple, there is a pericope titled the Widow's Offering wherein Jesus places himself near the spot where people come to render their offerings for the temple and its treasury and he comments on one of them. Several donate large sums given their wealthy status, but a poor widow leaves only two small copper coins. Jesus praises the widow because, proportionally speaking, the wealth of the rich will not be unduly diminished with their offerings, but the widow gave all she possessed, hence one hundred percent of her available funds. Thus she gave a proportionately greater amount than anyone else, even all of them combined (Mark 12:41–44; Luke 21:1–4, the Mark passage rendering greater detail). The take away is that she gave until it hurt, despite its being a tiny sum and likely the smallest denomination of currency present at the time, the tradition referring to it as the widow's mite ever since. In pointing this out, Jesus was attempting to impress upon his listeners that the amount we tithe (understood as a tenth) is not the main thing but how proportionally generous it is. The episode has memorialized hers, and we would do well to follow her example. Plus, God has bestowed gifts liberally, and we should do the same.

Definitely a tale worth telling and remembering, yet I have some comments. How was it, to begin with, that Jesus was aware of her financial status? He was not in Jerusalem enough to have made many contacts. Unless of course we play the omniscience card and claim that he knew everything there was to know anyway, given his divine nature. That is the go-to option for some, but as we mentioned last time, there are several passages which inform us differently and suggest that his was not as extensive as God the Father's. Plus, I find it interesting that, although perhaps implied, the text says nothing in the way of "and she will be blessed for her kindness," or "her giving will be rewarded and return to her manifold," for Jesus was present at the time to at least promise or at most produce such a blessing upon her, but did not, or if he did it was not recorded. Nor was he above blessing in a material way as in the abundant catch of fish after his resurrection demonstrates in John 21:1–14.

Sixthly and finally, a quirkiness on the part of our species is the following: when our character is evaluated or, say, our employment performance is up for review, the negative aspects psychologically far outweigh the positive and "dominate (or even overwhelm)" them.[1] We are much more inclined to take to heart the negative reviews over the positive even if the positives far outnumber the negatives. It is the negatives we tend to remember and which can eat away at us and even cultivate resentment and other ill feelings. We can't help it; we have become built this way. Yet through concerted effort we can mitigate the blows to our egos, insecurity, and self-esteem, and concentrate on the positives.

The situation is quite the opposite, however, in other cases. When one is informally experimenting, for example, on the extent to which one can forecast the result of coin tosses or dice throws, one can tend to admire one's abilities where hits outweigh even a greater number of misses, thereby deluding oneself. Or for those alleged psychics who are consulted by law enforcement to disclose the whereabouts of deceased persons, the information can tend to be vague and when one generates a hit after numerous misses one can convince oneself that "See, I really do have the gift," even though a more sober look at the track record can reveal otherwise.

What is it that makes us focus on the negatives in one set of circumstances but on the positives in another? Indeed, how is it that we think negatives tell the real story in one set of cases but the positives in the other? At the very least, we are inconsistent in this regard and can readily be swayed. Why has natural selection encouraged us to hold such distorted images of ourselves and our capacities? I ask once again, what accounts for such a shift in attention when it seems that both, one could imagine, should be geared toward the salutary in one's self-concept and where reflecting well on one's person would be the most desirable outcome? As the story goes in the recounting of orthodox evolutionary theory and history, the factors must either be advantageous for the leaving of offspring or be adaptively neutral. It is not readily apparent as to which is the case here.

1. Rosen and Royzman, "Negativity Bias," 298.

FRONT DOOR BLUES

I wish to regale you, dear reader, at this time with personal anecdotes. They stem from encounters with religious organizations which arrive at the front door to introduce themselves and their doctrines most eagerly to you without, I must hasten to add, invitation. Two groups in particular come to mind. The first is the Church of Jesus Christ of Latter Day Saints, otherwise known as the Mormons, flogging as they do the book by the same name. Two well-dressed gentlemen, always with shirt and tie and always male (evidently women cannot be trusted to engage in such a heady undertaking), came to the door and promoted their position in an attempt to prevail upon us the need to accept their scheme for ourselves, hence evangelism of their particular sort. I am always cordial and invite them in so as to converse while comfortably seated.

I then proceed to insist that they account for their foundational views, such as the authority conferred upon their scriptures, whereas they merely assume them without critical reflection and concentrate on their sacred texts as they stand. Plus, once they appeal to the founder of their movement, namely Joseph Smith and his experiences, I respond with the question as to why it is that anyone should place their confidence in the revelations given to him by an alleged angelic messenger instead of, say, Muhammed, who had analogous experiences much earlier, though not in specifics. Both received their messages from these other messengers. Hence, I propose, if this is what impresses their movement, then why not focus their attention on one who has priority in this respect? Why not, then, opt for Islam in place of a later or copycat version?

This appeared to hit home with the junior devotee who turned to his superior and asked, "Yeah, why is that?" The leader had no response and suggested embarrassingly that we end with prayer. This we did and, taking their leave, they never returned, surmising most likely that we were a threat to their brotherhood, plying them as we did with questions for which they could not provide an automatic pre-scripted rejoinder. I imagine they blacklisted our address. Too bad for them; we had coffee and cookies to offer.

On another series of occasions the Jehovah's Witnesses graced our domicile. As is our custom, we proposed to them that the sacred texts upon which they draw have not as yet been cleared for takeoff: the first order of business needs to be convincing us that their ceaseless quoting of Scripture is something that carries weight. They failed to draw us into the discussion by not first persuading us that their source of authority is also imposing for us, since it is not. We of course are playing devil's advocate for the same thing can be said for the Judeo-Christian tradition and its scriptures. This becomes an exercise in the philosophy of religion, some would say apologetics. Before we open this book of theirs, we need to lay the foundation that it is permanently binding on all people, and this they have not established. It is not obvious that their "good book" is worthy of devotion, for if it were we would all be followers, nor is it patently plain that their answers are directly related to the questions we have. Kindly ask us first as to what questions we come with rather than assume what these might be.

Their strategy was then to proceed with the counterpoint, yet another appeal to their scripture, namely the passage in 1 John 3:16 where "All scripture is valuable for instruction," etc. We informed them that, regrettably, this is self-referential: in philosophy it is what is known as question-begging and an appeal to authority where the move is made that these writings must be authoritative because the text claims they are, and the text could not be wrong, could it? It is like Muhammed Ali claiming that he is the greatest prize fighter of all time, for he would not lie about a thing like that, would he? The trouble is that he is not in a position to make that assessment about himself in an unbiased, non-prejudicial way, and therefore his testimony does not count. Anyone can say that s/he is the greatest, but one cannot be objective, least of all about oneself. It would better be left to later historians of boxing. Besides, John, the supposed writer of this passage, likely did not have his own treatise in mind when referring to what constitutes scripture. If so, he could perhaps be diagnosed with megalomania.

After repeated visits and attempts to convince us, bringing ever higher echelon types into the fray and bombarding us with their best salvos, they decided we were not sufficiently fertile ground and sought greener pastures elsewhere, though now perhaps not as galvanized in their own position. They hurriedly excused themselves and we have not seen them since.

Subsequent to our next relocation, perhaps they did not receive the memo to avoid us at all costs for leading their flock astray, but descended once again upon our front steps. This time there were two women! Take note Mormons! One was a supervisor and the other a trainee. The latter was silent as the former droned on about something of great significance to their ilk, though hardly making a ripple in our world. They insisted that God has but one name: Jehovah; all the rest are merely titles. Jehovah, however, is an English transliteration of the Hebrew YHWH, Hebrew not coming with vowels though it does have vowel points, a consideration about which they are either blissfully unaware or defiantly indifferent (if that were possible and not merely a contradiction of terms). Apparently, unbeknownst to them, the remainder of the world remains entirely untroubled about this matter.

Be that as it may, I explained to them, despite their having erected battlements against any naysayers, that the term El Shaddai is also a name, derived as it is from El, the name of the Canaanite divinity which the Jews adopted. The response on the part of the supervisor was, "Well, the Spanish can call God anything they want." At this I was dumbfounded, completely nonplussed. My reaction should have propelled me to declare, "That's right, the designation stems from the Spanish part of Palestine!" Yet my mind was not presently engaged with my mouth and gloriously failed to convert the thought into an audio component, whereupon they took their leave. They nimbly, deftly, and with pluck, though without aplomb, turned about and strode off, and by their actions made it abundantly clear that I was not a mission field open for their conquest. This left me thinking, "Hey, we're supposed to be doing that to you!" I virtually had my own door slammed in my own face! The ignominy of it all!

Should there be a moral to this recounting, it could be the following: Every position has its strengths and weaknesses and comes at a cost. To concentrate simply on the former is to be self-delusional at best and dishonest at worst and even to exhibit cult-like behavior. Growth occurs when one can be critical about one's own stance. I would like to capture an instance when one of these two groups visits and attempts to proselytize the other. The vigorous debates which could ensue, unless they excuse themselves from the ordeal, would be most entertaining.

For those which qualify, cultic organizations, as it happens, have determined that a sense of belonging on the part of their adherents trumps rationality. A message aimed at the heart persuades more so than does reasonableness. Some are being sold a bill of goods but are willing to sign an informal waiver stating not to bring the full force of their reason to bear on the group's doctrine in exchange for the assurance that they are part of the initiated and, in turn, are on the correct side or in the right camp.

If I may be indulged for a moment and allowed still another anecdote having nothing to do with religious groups, it might afford some comic relief. When we were about to relocate to a certain locale, we embarked upon a reconnaissance mission and sought a motel room so as to continue our search the following day. At one such establishment, we asked if we could first see the room in order to determine if it could meet our exacting standards. The proprietor, bearing an ethnicity not integral to the account, stated that it was not his policy to permit this but relented and gave us the key. We saw the room, were underwhelmed, and noticed that a light bulb had burned out. We returned to the front desk and made mention of this. He said that a Walmart was just down the road and we could obtain one there. We responded by stating that if he could not cough one up himself that we would seek competing accommodation, to which he ejaculated, "See why we do not let people see the room first?!" Once again a suitable and timely comeback did not present itself for me and we departed, leaving my wife and I bemused about his form of rationality, the futility of engaging him in reasoned debate, and all three of us disgruntled. The futility aspect is what connects this tale to the other three.

AS FOR MYSELF

The concept of the self has not always received applause and fanfare, Buddhists and David Hume as prominent examples. But to reiterate from and elaborate upon last time, and even before that a main theme in the previous work to it, I wish to recall the issue of the divisions of the person. To begin with, I understand the soul as the deposit of what the mind reasons, the psyche feels, and the will evaluates and decides upon. Think of Oscar Wilde's *The Picture of Dorian Gray*. The painting becomes a reflection of what was occurring in Gray's soul, the poor choices he was making in life, though unnoticeable externally by an onlooker.[1] This is the shape that the soul can take. Unlike Gray's portrait, however, humans are living souls (Gen 2:7) and not simply signposts or snapshots of otherwise invisible interior characteristics along life's way. These qualities might in fact rise to the surface, but they can be submerged as well.

Continuing on the topic of divisions, the passage in Heb 4:12 is informative: "For the word of the Lord is living and active. Sharper than any double-edged sword, it penetrates even to dividing soul and spirit, joints and marrow; it judges the thoughts and attitudes of the heart," which implies that soul and spirit *can* be divided as, at least nowadays in medical science, can joints and marrow. This provides additional fodder that humans are composites.

Recalling another topic from the former discussion, Sartre is correct in that our essences are made by our choices; he is incorrect when it comes to thinking that we are without one when we enter the world. Our essence is already in place, it is just without form at the time. Soul-formation occurs as we employ our will and reach decisions, and the cluster of choices has both an interior effect and also an exterior one. Sometimes it is claimed that the eyes are the windows of our soul, and we can determine by viewing them if a person's "lights" are bright or dark. It might not take long for our "quality-o-meters" to register when we are dealing with a person whose demeanor is gruff or pleasant, attractive and inviting or repellant. As Wilde would concur, our soul is reflected to some extent in our personal life, but as such the soul does not act and is not an agent; rather, it is acted upon. We

---

1. Wilde, *Collected Works*, 1–154.

shape it and act in accordance with it, until such time as we alter it with the decisions we reach, for better or for worse.

Catapulting us from the concept and identity of persons to the relations between them surfaces the issue of long-term relationships, what most have called marriages. One reason why many of them fail is that people tend to enter into them insisting, though not usually verbally, that their lives should remain the same, only that their partner become a useful accessory, one that could enhance their already existing regimen, not realizing that the partner probably feels the same about their own world and thereby registers a counter-insistence. When they discover that their lives are not augmented as anticipated, the partners become faulted for holding the other back from their interests. And they have the temerity to speak before being summoned. The frequency with which some persons change partners, as often perhaps as some people purchase a newer model vehicle, makes the relationship seem more like renting or leasing than marrying (leaving aside for the moment the actuality that all we possess in life is merely rented anyway). The proper mentality to adopt for a marriage to survive and thrive is that they are not to be taken as fifty-fifty relationships but one hundred-one hundred, that is, not disposable but all in.

WHICH LIVES MATTER?

Racism has a long history. Far too long. Ever since there were isolated cultures, separated by geographical features such as forests, rivers, lakes, and mountains, suspicion of nearby peoples was fed generationally concerning the clearly monstrous sorts who live on the other side (which is what M. Night Shyamalan's 2004 film *The Village* treated), thereby stoking the fires of the "us versus them" distinction. The unknown becomes the threatening, until such time as they become known, for then at least we know whom to vilify.

Historical documentation stemming from the ancient world and attesting to such division after the cultures came in contact include the biblical book of the Genesis myth—the first, though not the oldest, listed work in the Judeo-Christian canon's table of contents. In it we find a passage in which we can probably have substantial confidence, since there is little reason to suppose that the author had a hidden agenda to uphold. Concerning the episode of Joseph, the youngest son of Jacob, having been sold by his brothers as a slave to merchants heading to Egypt, Joseph needless to say had an initial downturn in his fortunes but rose in rank to become second only to Pharaoh. His brothers later travelled to Egypt in order to secure food during a time of famine. They encountered but did not recognize Joseph, he having grown from a seventeen year old shepherd (37:2) to a high-ranking official in this foreign nation.

Prior to his having revealed himself to his brothers, he hosted a meal for them, and it is here that we note an instance of discrimination. As intimated before, we are informed that the Egyptians ate separately from the Hebrews, since to do otherwise would have been contemptible to the former (43:32). The practice is stated as an aside, meaning matter-of-factly, describing the way things stood without denigrating the ones nursing the prejudice. It simply and casually reflected an observance to which both sides at the time had grown accustomed; so live with it. Given the accepted practice, it would then be best to focus on activities from which both sides could benefit, typically economic, so shrug it off and get on with the business of living.

But in the process, suspicion grows into undercurrents of hate, which in turn leads to exploitation and potentially aggression. This is precisely what occurred to the Hebrews generations later, as the text of the Exodus legend indicates. It would be salutary to think that if we all just become cosmopolitan in outlook, our biases would dissolve, but multiethnic does not automatically translate into appreciation or integration, nor does education seem to achieve the undoing of biases. In New York City, for example, Afro-Americans are removed from view into the ghettos of Harlem. Similar cases obtain for many other major centers, and it does not appear that there is an appetite on the part of some white folks to alter these circumstances. Same thing for North America and its Indigenous peoples. South Africa was so impressed specifically with the Canadian policy of leaving First Nations on reservations that they adopted the same strategy and called it apartheid.

While on the topic, I find myself being struck by the similarity between the OT book of Joshua and the plight of First Nations peoples in what is for us but not them the New World. According to the biblical text, the Israelites at God's command moved into a land not their own and thrust out the inhabitants by way of the extermination of each nation they came across. God's rationale was that if the children of Israel failed to do this, then they would be tempted to adopt the ways of the various peoples, namely the shameful practices of worshiping their false gods, and thus Israel would commit idolatry against God. The Israelites, having been given many victories at what they interpreted was God's hand, chronicled these exploits in OT books including two by that very name (Chronicles). As we know, the victors get to write the history.

So with the Europeans who sailed to a land not their own. They encountered the Indigenous peoples and deemed them to be brute primitive savages who needed to be saved from themselves. This paternalistic notion stemmed from the belief that God had sent the voyagers to convert the natives, and were that to fail, the recourse was exterminating or isolating them in parcels of land called reservations. It is a sad commentary that they had no reservations about doing so, believing they were doing God's work. Yet it was criminal to confine a nomadic people to one site. The Crusades spring to mind as another example, as does the policy in Tasmania of going from one end of the island to the other and wiping out the Aboriginals in their path in the conquerors' version of pest control. And as before, the vanquished did not get to write the history.

Only recently have we learned of the atrocities levelled against the invaders. The endeavor to fashion the natives into proper European culture was racist and pernicious. God must not have been proud. The brutal treatment of these folk has come to light. I wonder what kind of history they would write, and they will get the opportunity, for they have been given a voice and are availing themselves of it. The resulting picture will not be pretty for us.

When I contemplate how readily some nations can demonize others, like Serbs and Croats, whether or not in times of war, and foster a propaganda machine to feed it, as in the Third Reich, I get all choked up, and not of the favorable kind. Recent repulsive events surround the mistreatment of Black people at the hands of law enforcement, igniting the rise of the Black Lives Matter movement. Black people are justifiably incensed about the frequency with which some of their lives are lost in this way, largely by white people with firearms, and knees. Black people are definitely receiving the brunt of it, their treatment legitimately described as sub- or non-human. How unsettling it is to witness how much better some people treat their pets and automobiles than the Black community in particular.

Yet this is not the only repugnant event to occur of late. As a result of certain figures referring to the agent of the contemporary pandemic as the "China virus," the detestable practice of assaults on East Asians is fuelled, for the crime of simply being East Asian and the belief that all of them were personally responsible for the onset of the coronavirus. I have viewed multiple disturbing CCTV video clips of East Asians being persecuted, most involving battery, and one of a proprietor of a convenience store being pepper sprayed, making these acts hate crimes. Now here's the kicker: the assailants were sometimes Black! You do not augment your case for fair and equal treatment or propel your cause toward curtailing racism when some of your own actually perpetrate it. They are demanding anti-racism while at the same time are unwilling to extend it. Evidently, some of them know how to be racist too, and are quite well adept at it. All sides, apparently, must shoulder some of the blame.

Returning to the themes of South Africa and film, I am reminded of the South African film "District 9," where extra-terrestrials found themselves stranded on Earth and, like the Black people there until fairly recently, were confined to the squalor of an actual camp named District 9. The aliens had become a nuisance and an eyesore, even compelling a Black man to declare, "I wish they would all go back to where they came from." The use of irony was very effective, though it was lost on that character.

To be quite frank, a complete and thoroughgoing toleration is not something one should or even could strive for, since there are certain acts which ought not to be tolerated on either side and toleration can engender an attitude of indifference where certain groups are still considered as lacking the respect due to others. None of us is immune to this. Not least of which is the onerous fact that omni-tolerance is not a possibility, for we might then find ourselves in the unenviable position of needing to tolerate intolerance. Yet in order not to be disrespectful and in the interests of political correctness, itself not viewed as essential but as a necessary inconvenience, we believe we are being civil when we extend a metaphorical hand not to welcome but to keep at arm's length. Those campaigning for political correctness may be well-intentioned, but it might not have the desired effect if the heart is not in it. Better perhaps to aspire to acceptance which can lead to compassion, rather than tolerance, which might not.

Now you might be asking me, your basic standard white guy, if I do not harbor some racist sentiments of my own. My response is, of course I do. I state with all candor that I do not have confidence in some of the people of my kind doing right by some people not of my kind. We are sometimes the scourge and have been party to the subjugation of other peoples with our colonialisms and empires. The offending events on the part of white people are simply too plentiful for me to overlook. There is a long tradition of presumed white superiority, even supremacy. And people in those groups are about as easy to get through to as Holocaust deniers. There are despicable, ghastly acts and policies that my kind has concocted and underwritten, making it farcical to suggest that white people are supreme. Other adjectives come to mind if you use your imagination.

Aside from atrocities leveled against race, gender, or sexual orientation, keep in mind that most who become mass murderers and bombers on North American soil tend to be white. Face it, some white people are easily radicalized. And as the etiquette of mass murder demands, those who become snipers are equal opportunity killers. Think of the incident at a Las Vegas concert and the nearby hotel. Oftentimes the victims are eulogized as having been pillars of the community; the assailants usually not. Scarcely are the perpetrators ever accused of being a credit to their kind.

Furthermore, if one is intent on removing statues of famous figures were they to have been slave-owners, then one should consistently also take down all images of Charles Darwin, for he, like most Brits at the time, were racist. It is easy to denigrate former and foreign populations for what they should have known and I am not here to defend them, but we having become enlightened ought to prevent additional injustices from occurring and hence concentrate on the present while avoiding the ignorance of the past. It all boils down to this: the best persons in any culture are stellar individuals, exemplars, and role models; we would do well to emulate them. I have more to say about these matters, but that would employ language, contrary to the New York Times, that would not be "fit to print".

Plus, on two extremes, what I can say is law enforcement brutality needs to stop and a single act should not define an individual, together with accusation is not the same as guilt, for everyone first deserves his or her day in court. Also, a person should not be treated as a criminal if s/he has been cleared of all charges, for exoneration in court ought to translate into society. I have in mind a personal friend for whom this courtesy was not extended. But in an effort to end on a high note, we would all do well to strive for inclusivity, for it is in our power and not beyond our reach; we need only the will.

PRIDE AND PREJUDICE

When critics of religion refer to it, they tend to paint it with the same brush strokes as though Christianity, the one most specifically yet not solely in their cross-hairs, were a single monolithic stance. This is inaccurate, for not all is well in the religious home. By this I do not intend its fragmentation into a multitude of denominations, though this is also a concern, but the mindset of some of those within them. We do not all see eye to eye, that is part of the human condition not only in religion but in many social settings—just witness politicians at play for a rather onerous spectacle of how not to behave. Instead, I have in mind the scope of what some religionists presume they must be the gatekeepers. Or, to employ a different analogy, they have been found to thrust themselves into the role of referees over the type of subject matter allowed to be given a hearing.

Intellectual freedom, the alleged property of college and university faculty (though it does not exist widely even there—just try to get funding for a politically incorrect topic such as which race performs best at mathematics) is under a tight rein in some churches and institutions of higher learning. The attempt to have a civil round table discussion on practical themes like gays, lesbians or women in the ministry, or theoretical ones such as the extent to which Jesus must be understood as both fully human and divine, could very well be met with hostility, and not of the good kind. And when pressed as to why these issues should be the ones the self-appointed enforcers are to be trustees over, other than the knee-jerk use of biblical prooftexts, little is offered in addition to hard core intuition.

Let's take an example. One difficulty I have with some of my religionist colleagues is the following. In the past generation there has been a spate of writings geared toward debunking religion. Authors including Richard Dawkins, the late Christopher Hitchens, Sam Harris, and Daniel Dennett have produced works (as well as lined their pockets) that amount to screed about how we should all grow up and dispense with religion, for it is a veritable blight on the human and intellectual landscape. Religion simply cannot be accurate, they claim, because it all stems from a non-empirical, read unscientific, basis.[1] Much of the argumentation is poorly constructed and

---

1. Haught, *God*, x-xii; Hedges, *When Atheism*, 53–53; Hitchens, *God*, 282–83; Ward, *Is Religion*, 85–86.

does not lead to the conclusions they anticipate. What is worse, it may very well be unfalsifiable and thereby rendered vacuous.

Armed as they believe they are with science in their corner and at their disposal, they are at the same time unaware that they have closed themselves off from refutation, which is a bad thing in science. Theories must be vulnerable to falsification from competing evidence in order to qualify as science. And it keeps scientists humble. There must come a point where a hypothesis can, at least theoretically, not withstand the counter-evidence with which it might conceivably be confronted. If not, if nothing is permitted to count as undermining evidence, then the theory has disqualified itself from standing as a legitimate scientific claim and becomes but propaganda. When it comes right down to it, as I intimated last time, all art and writing is propaganda, it cannot help but be this way, but what we should attempt to avoid as best we can is contributing to it. The above-mentioned authors have, at least at times, committed this error and have therefore overstepped their bounds and, as a result, we do not need to take them seriously.

This is ground which has likely been covered already, and if not, it should have. But this is not my main concern here. Some religionists would simply refer to these works as nursing an anti-supernaturalistic bias, and harboring any prejudice is not something proper science cultivates. Those who have it do a disservice to science, for such a personal preference has no place in it. So much is clear and uncontroversial. What all this means is that the authentic science which anti-religion authors uphold is not reflected in their own work.

Those religionists have a point, but only thus far. They are correct in that such activity rejects the very spirit of discovery which is the touchstone of science, but have curiously veered off-course into something unscientific and anti-metaphysical of sorts on their own. True, anti-religion authors have denied the empiricism which they claim to champion by concentrating on religion's perceived evil (not altogether undeserved) without countenancing its good (there is some of that too), thereby revealing that they have become the strikingly opinionated few with lab coats, bullying a caricature of religion which thoughtful religionists themselves are equally disinclined to accept and defend. But anyone can spout opinion. It does not require scientific credentials to opine on a subject not one's own.

Since it is futile to offer answers to those not asking questions, the ones who are still seeking and working on becoming self-critical and non-prejudicial mark the extent of the religious company I keep, for the religionists above might very well fall into the same trap as the anti-religionists they criticize. They rightfully vilify the "New Atheism" as it is called (which is at least as old as theism itself and hence not new, it just incorporates recent science), while at the same time betray their own shortcomings. There is also a line of inquiry against which they have shown themselves to wear blinders. Whereas the divine for them warrants inclusion as an area worthy of study and an acceptable academic pursuit, they do not allow, say, parapsychology into the mix, as if to say, "God? Well, sure, let's pay attention to this particular unprovable metaphysical notion, but ESP? Preposterous!" And with this they have demonstrated themselves as woefully inconsistent.

Despite one's position on the issue, and one could readily interject a different one (like whether they tellingly bear an anti-*natural* bias when it comes to the questions of the onset of life, mind, and consciousness, as outlined previously), do they believe that "Here be dragons" after all, a scholarly bridge we ought not cross? It seems humans tend to allow for themselves what they do not permit in others, and this does not speak well of us. On the one hand they appeal to a certain segment of the population which they claim should be more open to things supernatural, and on the other they close the gate in addition to their minds to the possibility of something else metaphysical—that some humans might perhaps come with an exceptional ability like telepathy or clairvoyance. Regrettably, their minds are selectively open and closed, and the selection process does not appear to suggest equitability.

As it happens, parapsychology was granted albeit reluctant affiliate status in the American Association for the Advancement of Science (AAAS) in 1969 (over half a century if someone cares to do the arithmetic), and has at least had research centers at Duke and Princeton Universities, and programs up to and including a doctorate have at least been offered at the University of Edinburgh. Its scientific character is and has been recognized by some, such as the philosopher C.D. Broad,[2] and it ought to be respected as a rigorous psychological sub-discipline, experimentation mostly having become refined so as to eliminate staging on the part of subjects.

I personally hold but marginal emotional investment in this, as some would charge, peripheral- or pseudo-science, yet it can be used as a helpful test case, especially if one is in search of a handy scapegoat. The Bible, it must be noted, does not help us significantly for the specific capacities mentioned here. Instead, King Saul, as in the 1 Samuel 28 account, approached the witch of Endor to act as a medium to call up the shade of Samuel so as to garner advice on what was to become Saul's final and life-ending battle. This is an example of spiritualism, expressly forbidden by the Mosaic Code, and is not what parapsychology properly trades in, for the former is more the province of the paranormal.

---

2. Broad, *Lectures*.

We are also warned not to delve into matters too wonderful for us (Ps 131:1), but are not informed as to what these might be until we arrive at Ps 139:6, where the matters involve God's extensive knowledge of humans, and Job 42:3b, which refers to the sum of Job's responses to his "friends" dealing with God's counsel, ways, and economy. Even the Jesus of the gospel text had his clairvoyant moments: Matt 17:27 has Jesus stating that "so as not to offend them [the temple tax collectors], go to the lake and throw out your line. Take the first fish you catch; open its mouth and you will find a four-drachma coin. Take it and give it to them for my tax and yours." (My comment on this passage is since when was Jesus concerned about offending the authorities? The answer: he was just being a responsible Jew.)

Additionally, Jesus displayed telepathic episodes during the times he was said to have known the thoughts and hearts of others, and also exhibited precognitive ones when speaking of the destruction of the temple (Matt 24:2), and about his resurrection (Matt 17:2; 27:63; Mark 8:31; 9:2; Luke 18:33). In fairness, though, we should admit that given the probable dates of the drafting of those gospels, these sayings were likely placed in the mouth of Jesus well after the fact. But we are for the moment simply presenting the text as it stands.

These instances, occurring as they do in the scriptures, must be assessed as positive by those holding to biblical authority. Unacknowledged by the religionists I have in view, then, they are therefore in favor of ESP, for Jesus expressed it (and they also believe in a Ghost, only the Holy one). Now they might declare that no one else but he, being Messiah, must come with some attendant perks and was privy to them, but the point remains that God bestowed these powers upon a human, and subsequent to the time the Spirit was outpoured, humans were given some abilities not widely practised or even previously available. I recommend consulting for yourself the treatment in the book of Acts to this end: from speaking in those tongues truly foreign to the speaker but understandable to some hearers (2:1–11); to Peter's healing of infirmities (3:1–10), sometimes not even by touching but simply via casting a shadow upon those seeking healing (5:12–16); to Paul's healings at times merely through the touching of a piece of cloth that Paul had also touched (19:11–12); to Paul's outright raising of the dead (20:7–12).

Such properties have not been alien to some humans regardless of how they came by way of them (a topic I have addressed elsewhere). A tight bond, for example, can form between persons, as with couples or mother and child, where one can know when the other is in distress, and Christians are informed as to the provenance of "every good gift," namely the divinity itself (Jas 1:17). What then is the nature of the subject matter making it taboo in the eyes of some religionists? If parlor tricks can now be ruled out, then what else stands in the way? And if these phenomena are to be ruled out and successfully avoided, it must be on grounds other than subjective personal preference.

Thus whereas the religionists in question reject an anti-supernaturalistic bias, some appear to retain an anti-metaphysical one in the human realm. Or, should it be a natural capacity, it could then be termed an anti-exceptional-ability bias. My request to the religious naysayers with their territory to protect is for them to provide me with a rationale as to why I should not take the further step and label them as hypocritical, *for I am open to one*, but it must be more than simply, "why it's of the devil," for then so might hypocrisy.

THE DOORS OF PERSPECTIVE

Following on the heels of the end of the immediately previous topic, I undertake this next segment with trepidation. I realize that I am not speaking to all readers here, since they might simply bracket the question, yet I remain convinced that it is worth asking. Among the issues I struggle with includes what in Job 1:6 means the accuser, for I do not wish even to dignify it by using its proper name, but rather concentrate on what it allegedly does. We are informed that in addition to God and God's servants or heavenly court, implying that God is a king, there are principalities and powers, spiritual forces in high places (Eph 6:12 KJV; the NRSV employing "cosmic powers" and the NIV "heavenly realms") that or who militate against divine efforts to expand God's kingdom and which God permits to persist until such time as God's rule is fully and visibly instituted. What troubles me is the pattern by which we either wittingly or unwittingly become enablers of such negative forces.

By Roman Catholic accounts, exorcisms of demonic spirits having taken possession of individuals are on the increase, as is the need for and preparation of these spiritual therapists, so those who can place exorcist onto their curriculum vitae can expect to be considered for gainful employment.[1] But what are the instances in which their training is called for and are they consistent? Some reports claim that the use of Ouija boards (the translation from the French and German, respectively, is the straightforward "yes yes") opens the door to these elements,[2] not so much by the casual use of the dabbler, but by sustained practice over extended periods. If true, this entails that such forces can be unleashed without having been summoned and the practitioners need not even invoke the names of these spirits or request their presence in order for them to surface and carry out their destructive work. I find this odd, for I had suspected that some verbal and active formulae were required for this to occur. Evidently not, and since not, we can be unknowing participants in their appearance.

Strange, to reiterate, for there are also Satanist enclaves that knowingly invoke spirits through the performance of specific rituals, and, assuming they are effective, do not necessarily imperil themselves in the process in

1. Mariani, "American Exorcisms."
2. Mariani, "American Exorcisms."

the way that those who are possessed do, for the latter face the miserable prospect of being taken over against their will by forces not their own. Yet Satanists can readily walk among us undetected. Which prompts me to ask what the rules of this game are, for sometimes these forces manifest unawares to us and at other times they do not or must be invited. But what constitutes the invitation?

Humans can engage in any number of injurious activities, many perhaps propelled by dark forces. Anything, for example, can become addictive if we happen to obsess about it. Some people are gamblers, though not necessarily to their own detriment, unless it becomes out of control. Others are ensnared by erotica, which can result in both deleterious personal and interpersonal expressions. But not always. Some enjoy fine cuisine; others are gluttonous. Anything, I repeat, can become distorted, and this seems to be the program of the dark side—to pervert anything of value. Yet this affects not only the positive but what is already negative by amplifying it. Anger can turn into rage, dissension can become division, and resentment animosity, to name a few. So what are the parameters within which we can expect such occurrences to arise?

There are certain activities which might be considered prime targets for the chaotic. Seances, which seek to raise and communicate with the spirits of, say, the dearly departed, are a case in point. We cannot be certain that the doors we are knocking on here do not open onto dark spirits who are simply counterfeiting events. We have no way of telling whether this force is Uncle Nick or the demonic and might best be left alone, since we could be asking for trouble. And when it comes to amplifying negativity, we can be caught in a descending spiral if we were to allow negative thoughts to run amok. These are easy traps to fall into but are not always easy to climb out of, especially if we, for instance, act out our aggression toward someone else.

Science enters the fray and asserts that nature (specifically one's genetic complement or genome) can predispose one into a certain behavior pattern, though it requires nurturing (a particular environment) for it to become expressed or fanned into flame. Or since we have broached the topic of behavior, it can be said that this is the department of a psychiatrist or psychologist. That would be fine if the theme of the demonic is regarded as a legitimate line of inquiry, for ordinarily it is not. And when it is not, the afflicted person if of sufficient severity can be assessed as inherently evil, should even that category be perceived as worth pursuing. Hence we are left with the difficulty of recognizing that there can be destructive elements at work, only we do not know if they stem internally to the person or externally. And if the latter, who or what permitted them entrance? A strict correspondence of activity A leading to behavior B (conveniently these terms begin with those very letters) originating through spirit C (oh well, two out of three), cannot be charted in a straightforward manner, entailing that some who play with fire get burned and others do not. Perhaps this is one of the things that are "too wonderful for [us] to know" (Job 42:3b), for there are some things to be wary of.

But can these forces at least do us the courtesy of being as orderly, predictable, repeatable and calculable as are all the physical forces we investigate? Or again, can spiritual forces not be a little bit more scientific? Is this asking too much? But then what do you expect from non-law-abiding spirits? In the prior section we concentrated on positive aspects and what might be regarded as good gifts of God, working as they do for the benefit of humans if treated appropriately, for these too can be obsessed over. Here we focus our attention on the negative, which by its very nature is inimical to the divine, though angels too can be sent out to perform destructive acts (Ps 78:49), purportedly of reprisal.

Nevertheless, I cannot leave this section as it stands, but must pursue another line of inquiry prior to ending it and will do so in the following manner. In addition to the scriptures presenting varying views on certain themes, there is also a different complexion to discuss, namely an outright movement or shift in focus from one viewpoint to albeit related others, in the plural. One way this comes to light is found in the OT where alternative stances are taken in progressively later stages in Jewish literary output. The first is the conservative understanding of the way the world works, or more precisely the system or economy which the divinity has set up, as operating according to a one to one correspondence of acts perpetrated and just deserts earned for them. This can be located in the wisdom books of poetry, specifically the Psalms and Proverbs. In them one can find the sentiment that our acts will recoil on us, essentially if we have sown the good or the bad, then that is what we will reap, and this will be dispensed visibly in our lifetime. Examples of this are seen in Prov 11:18 and 22:8.

The language employed here is often in the form of the wicked perishing and the righteous flourishing and thriving. The rewards for the latter tend to occur in the shape of what was most prized for this life in the Ancient Near East, namely long life and inheriting the land. Next in line, marking the first transition, comes the book of Job where the righteous protagonist has been visited by personal affliction exacted upon his offspring, possessions, and physical body. His four "friends," and I use the term loosely since they are hardly a source of comfort, who mercilessly torment him verbally even further, represent the voices of the aforementioned conservative mindset. Their logic runs in the reverse direction from effect to cause in stating that as Job is the recipient of affliction then this must be the result of wrongdoing on his part, instances including Job 4:8, but it is not too late to change his ways, for God will then turn the curses into blessings. Yet Job is adamant that he has not transgressed and makes his case of complaint known to both friends and God. It can be put that way if one will allow this homonym to describe it: Job proclaimed his innocence and he was in a sense.

At the end of the work, Job is vindicated by the God of the whirlwind and the friends are not, for those who erroneously reflect the former position require sacrifices of atonement to be made for them (42:7–9). Oddly enough, the fourth friend, Elihu, was left out of the picture. What is not resolved, though, is Job's question as to why this has happened to him in the first place and, more broadly, why there is evil in the world. The thrust of the divinity's reaction is confined to the counter-question as to who are you to cast aspersions on my economy? The answer which the readers are left with is that the accuser, and this is what connects this subsection to the previous one, instigates the offense and God at least temporarily permits it. No resolution is offered, though, as to why God grants such permission.

The tone alters once we reach the book of Ecclesiastes, and with each successive work we are moving to the more chronologically recent in literary output. Ecclesiastes, and one can determine how much this work has been formative in my own thinking with the number of times I refer to it, provides a corrective to the initial position by declaring flatly that, opposed perhaps to the way we would assume or prefer it to be, the righteous die before their time and the wicked prosper and enjoy long life (7:15). This flies in the face of the simplistic one to one formula, since there are many instances to the contrary. The bottom line in this work seems to be that "the same fate" awaits both the righteous and the wicked, for God will bring all of them into judgment (2:14b; 3:17a; 6:6b; 9:3; 11:9; 12:14), thereby redressing all injustices. The implication is that it will occur in this life, yet then "a righteous man perishing in his righteousness" and the righteous and wicked receiving what the other has earned (8:14) would not appear to fulfil the intent. Thus on the one hand we are assured that all things will be made right, and on the other that a single lifetime may be insufficient to accomplish it.

The final instalment comes to us in the second half of the book of Daniel, which might be the most recent writing in the OT (with the possible exception of Esther), the second half coming significantly later than the first. In the twelfth and final chapter, we are informed that there is a book in which the names of the righteous are written, and "Multitudes who sleep in the dust of the earth will awake: some to everlasting life, others to shame and everlasting contempt" (verse 2). The genius of this literature is that the grave is not the end, but the time or era in which everything will be made right is not in this life but in an afterlife to which many will awaken. Hence there we finally have it: evil might very well have a mysterious origin, but it is fueled by the accuser and will be addressed and dealt with partially in this life and fully in the next, since this life does not provide us with either the timeframe or the proper arena within which the eradication of evil will be completed. At least in this instance, then, there seems to be a development to the biblical message as time proceeds in the direction of greater realism while at the same time greater hope for that which we cannot as yet perceive.

"IF YOU BUILD IT," THEY MIGHT STAY AWAY

There are multiple themes which not only capture my interest but will not allow me repose so as to leave them alone; I thus resign myself to "rolling with it" or "going with the flow." One topic addresses the artifacts left behind by ancient peoples. There are those which derive from the Paleolithic Era (the Old Stone Age), the Mesolithic (the Middle Stone Age), and the Neolithic (the New Stone Age). What occurred within the Old was the last Ice Age, which was marked by a series of crests and troughs,[1] the final stage of which is termed the Last Glacial Maximum, stretching from about twenty thousand to twelve thousand years ago and began to retreat subsequent to that time.[2] Unfortunately, of note is the disconcerting observation that few authors appear to agree on the dating of these eras. If we may hazard a mean of the figures involved, then the Paleolithic debuts from approximately forty-five thousand to ten thousand[3] years ago, the Mesolithic from ten thousand to eight thousand years ago, and the Neolithic from eight thousand to five thousand years ago.

Prior to these eras, humans became anatomically modern at roughly one hundred thousand years ago, and behaviorally modern at about fifty thousand years ago,[4] conveniently and perhaps even suspiciously conforming to fifty thousand year increments, and around this time or up to ten thousand years later was what is known as the Great Leap Forward,[5] when humans embarked upon, among other ventures, a religious quest for the eternal verities. Nevertheless, the issue for this segment concentrates on the stark difference between building projects at the end of the Paleolithic and the beginning of the Mesolithic. Beforehand there is scant evidence in the way of architectural structures to be found through archaeological investigation, only those items that can weather the meteorological onslaughts from that time to this, such as arrow- or spearheads and bones from burial sites,

1. Wade, *Before*, 29.
2. Wade, *Before*, 103–4; Fagan, *Cro-Magnon*, 32.
3. Wade, *Before*, 101.
4. Wade, *Before*, 8.
5. Ehrlich, *Human*, 159.

though pollen[6] and assorted otherwise ordinarily biodegradable substances can also be (literally) unearthed forensically, nomadic peoples clearly not investing much effort into the immobile.

After the Last Glacial Maximum (LGM) enters the scene and then exits from it, there is a veritable welter of structures, some of which defying our efforts to uncover how they could have been erected. What was it, then, about humans that was clicked on or sparked and fanned into flame (or thawed) rendering to them the interest as well as the ability to pursue and complete these ends? In addition to all the threats humans confronted prior to the LGM was then compounded by the advancing ice sheets. Were there humans then who thought to themselves, perhaps because the concepts had not as yet been translated into linguistic forms of expression, "*Homo sapien/* Man/Dude (check appropriate box), when this uninvited wall of water-grabbing insolence finally bids a retreat (it will retreat won't it?)," cogitations quite *avant garde* for its time, tapping as it does into the notion of a water cycle, "I and anyone else I can recruit will build, what should I call them, structures in direct defiance of the edifice immediately before us."

Kudos on them for their initiative which, regrettably, they were unable to enact, despite being incensed, which might usually catapult thought and ideas into action. The ice outlived them. Yet as it left and the water cycle resumed in earnest, there came a flowering of engineering seeds. What accounts for this? Endeavors such as temples and monuments became widespread whereas previously there were hardly any to be found. We came of architectural age. I am reticent to claim that it was due to cabin fever. Naturally, the transition from nomadic to a sedentary and then agricultural lifestyle[7] roughly eleven thousand years ago prompted humans to contemplate construction programs given the importance granted to some type of authority structure in a centralized location, but the issue is that structures became global in extent within a relatively short span of time.

The second theme I have in mind concerns both building and rebuilding. As for the former, when the children of Israel exchanged their wandering nomadic lifestyle for a sedentary one, as the God of the text commanded them to displace, even exterminate, the contemporary inhabitants of the land of Canaan, occupy their land, and reap what they had not sown, the divinity selected Jerusalem to be the "holy headquarters" and ultimately instructed the Temple to be located there (1 Kgs 5:3–5). Once settled in their geographical neighborhood, they were urged by the deity, in an about face from the initial extermination policy, to extend a charitable hand to

---

6. Fagan, *Cro-Magnon*, 55, 107; Stringer, *Lone*, 154.
7. Wade, *Before*, 125.

the foreigner in the land with whom they came in contact, from the rationale that they too were aliens in the land of Egypt and eventually were mistreated. The Israelites were to reflect the care of God now that they were in a position of authority over those displaced. Their Torah even contained legal injunctions for equal treatment of foreigners (Lev 19:33–34; Deut 10:18–19). Forward thinking, if not initially salutary for the recipients.

The following chapters in Israel's logbook involved the mentality of engaging in events and activities without discretion in direct opposition to the divinity's intent, thereby forcing God's hand to expel them from the land they had morally polluted. So off to Babylon they were sent by their captors. Jerusalem along with its wall and the Temple in which they banked their hope were demolished and looted. While in exile, the Jews reflected on how the deity could have separated them from what was for them the center of the world. They came to realize that the fault was their own and their forebears who misplaced their trust from God to what God had commissioned. Along with this evolved the view that God was a global and not a tribal or regional divinity. Once there was a suitable regime change in what became Persia, a sympathetic monarch opened the way for the Jews to return to their God-given homeland, as they saw it.

Those who availed themselves of the offer seemed to be the conservative contingent, however. Here is the reason why this appears to be the case. Those who returned endeavored to reignite the need to distance themselves from the surrounding cultures, since they believed this was the source of their downfall in the first place. Hence they were stoked with the urgency to vilify the other inhabitants of the land and perceive that the latter's impurity in turn made the Jewish people impure, were they to adopt alien practices, leading to a renewed surge in abandoning anything foreign. They ended up loving themselves and the statutes of their law more so than their neighbors, in contravention of that very law (Lev 19:18). They rejected the assistance of foreigners in rebuilding the wall in what was not unexpectedly to be called the Second Temple Period, and they sent away their foreign wives and children. This no longer looks like caring, and the books of Ezra and Nehemiah outline this right-wing movement and policies.

They thought they had reverted back to the basics, as in a more modern back-to-the-Bible movement. One can well imagine the fervency with which they sought to correct previous missteps, but the stepping went overboard. In sending foreign wives and children away, they were ultimately sending them to their deaths, since those wives and children would not be welcomed back by their erstwhile nationality either. They became people without a homeland, something the Jews themselves were also later to experience, on the order of millennia. This was a regrettable move on the part of the conservative Jewish constituency. They had an opportunity to build a bridge but did not avail themselves of it and instead (quite literally) (re-)built a wall.

With this we now launch into the second part of the study, where we will revisit the topic of ancient structures, and in the conclusion the wall-builders.

PART 2

Science

THE DOCTOR IS OUT, AND SO IS THE JURY

As Lewis Thomas has informed us, medicine is "the youngest science."[1] In times past, it was considered more of an art than a science, which is why old buildings dedicated to health professions were named "medical arts buildings" (where the "u" in buildings was shaped like a "v"—an acceptable early practice but regarded as unacceptable in the game of "Scrabble:" sadly they are not interchangeable; oh, the injustice of it). Then as now, and due to the complexity of our bodies, there is not a straightforward one-to-one correspondence between diagnosed ailment and prescribed pharmaceuticals. Sometimes it comes down to a matter of trial and error: "Let's try this. That didn't work? Well then let's try this. That works but it produces adverse reactions? Well then let's try this. That's a good start? Well then let's adjust the dosage to determine how much you can tolerate before the issue is resolved. That relieves the problem but the side effects need attention? Well then let's add this to counteract them." And so on. The remedy is believed mostly to be the strategy of throwing chemistry at an illness in order to correct faulty biochemistry. In extreme cases, where both physician and patient are at a loss as to how to proceed, perhaps due to a novel medical phenomenon, this approach can have similarity with what looks like a crap shoot.

Another youthful feature of medicine is the use of antibiotics. First used in wide measure in the Second World War so as to protect the armed forces from contracting maladies such as malaria, or to help troops recover from them, regrettably we are now finding their overuse to yield a law of diminishing returns. Nevertheless, a fascinating aspect of the Bible is its ahead-of-its-time response to microbial agents, although those ancient folks lacked any background knowledge of them, pre-Pasteurian as they were. These are mostly confined to "the third book of Moses," otherwise known as Leviticus, particularly chapters thirteen to fifteen. Consider these passages: "When anyone has a swelling or a rash or a bright spot on [the] skin," they knew enough or were allegedly given sufficient information from on high so as to diagnose the condition as a potential "infectious skin disease" (13:1–2). They knew nothing of the causes, though were well aware

---

1. Thomas, *Youngest*.

of the effects. Instructions apparently were delivered from God in order to "divine" how to properly determine their severity.

The priest was called upon to act as an amateur medical practitioner and "examine the sore" (13:3). If it presented in one way, a certain therapeutical tactic was to be taken, such as self-isolation (sound familiar?). The issue seemed to turn on the depth of the sore in the skin, whether it is more than skin deep and if it has spread. If not, "it is only a rash" (13:6); if so, "it is an infectious skin disease" requiring further isolation. Should there be "raw flesh in the swelling, [then] it is a chronic skin disease" (13:10–11), meaning, you guessed it, even more isolation. It is noteworthy that at least at their time there were no herbal remedies applied to the condition according to the accounts, although these measures probably had a long history; yet no reference is made to them in the text. Additional manifesting conditions include boils, burns, scars, itches, spots, and discolored hair; and in instances of the latter, the surrounding skin is to be shaved (13:33).

Perhaps most strikingly, such ailments are not restricted to visible human anatomy, but also have sartorial and architectural referents. Clothing of wool, linen, or leather may exhibit signs of mildew, which, if spreading, must result in the burning of the material, since the offending agent is described as destructive (13:51–52). Should it not be spreading, it must be washed (13:53–58). Cleansing involved administering oil, the blood of "live clean birds," as well as guilt, burnt, and grain offerings (chapter 14). Alarmingly, the text even states that God might actually *place* "a spreading mildew in a house" (14:34).

Once diagnosed as such by a priest, the contents of the home must be removed (14:36). If a "greenish or reddish depression" is seen "to be deeper than the surface of the wall," (14:37), then the house is to be barred for one week (14:38). At the end of this time, "If the mildew has spread," then "the contaminated stones [are to] be torn out and thrown into an unclean place outside the town" (14:39–40). The interior walls must then be scraped and the scrapings dumped in the same site allocated for the stones (14:41). The removed stones are to be replaced with others along with the application of "new clay and plaster." Should the mildew return and spread, the house "must be torn down" together with its timbers (14:45); if "the mildew has not spread," atonement must be made for the house by way of purification ritual observances (14:48–53). Sounds like the house was a bad moral agent.

Moreover, when persons have a discharge or a woman her monthly cycle, rituals are also performed so as to deliver them from uncleanness and make them ceremonially clean (chapter 15). Careful "attention to God's commands and . . . decrees" will result in God not bringing upon the people the same diseases God exacted upon the Egyptians (Exod 15:26), God being the one who also brings healing. These measures appear both primitive and modern at the same time.

All this talk about maladies and recovery from them seems ahead of its time. Even prior to the onset of the era of apparatus such as microscopes, the discussion sounds positively contemporary aside from the ritual aspect. The same, though, cannot be claimed for mental illnesses, admittedly more difficult to diagnose in ways other than merely "he has a demon." Israel's first king, namely Saul, is one such figure who suffered from a psychological dysfunction which the ancients, in typical pre-scientific fashion, attributed to the work of malicious spirits.

Nor is this knee-jerk reaction confined to pre-medical personnel. When radios were invented, some people surmised that the demonic resided in these wooden boxes in an effort to explain how they operated, or as they would have portrayed it, "by what manner of sorcery is this?" Hence one does not need to be ancient in order to reach those conclusions. Nevertheless, King Saul went against God's intentions, catapulting him into the unenviable position of standing at odds with God, who became disappointed with Saul's decisions, "was grieved that [God] had made Saul king over Israel," and ultimately rejected Saul as king (1 Sam 15). Saul then experienced a marked downturn in his equanimity and suffered a breakdown of sorts, perhaps partially brought on by this rejection but also the knowledge that he would be replaced by David as king and the jealousy this generated in him.

Researchers have commented on Saul's condition, diagnosing him variously as having suffered from, firstly, more than simply depression, and then the form of paranoia that does not prompt people to withdraw from others, but which lashes out in aggression,[2] evident when Saul attempted to pin his son Jonathan to the wall with a spear (1 Sam 20:33) and the same to David twice (1 Sam 18:10–11). According to the biblical evidence, Saul experienced psychological challenges, orchestrated by a deleterious spirit unleashed from the heavenly realms, for which the psychiatric world reserves several foreboding terms. Yet we cannot be certain, since the description we are given in the accounts are, contrary to the Leviticus example, entirely bereft of medical information.

Given the absence of usable scientific evidence, the ancients were left to employ their default response, specifically the influence of injurious spirits. What is remarkable about this instance is that these alleged entities are claimed to have originated with God for purposes of tormenting Saul (1 Sam 16:14–23; 18:10; 19:9). The benevolent Spirit *of* the Lord was replaced by a malevolent spirit *from* the Lord. When all else fails, turn to the demonic

---

2. Stein, "Case," 212.

as the cause, only the difference on this occasion was that this nefarious spirit was staggeringly in God's employ. Seemingly, this was a spirit commissioned to wreak havoc, but this sounds more like the Hindu god Shiva's method of operation as both creator and destroyer. Saul did find at least temporary relief through David and his skillful harp playing (1 Sam 16:23). Music can indeed, then as now, calm the savage breast (and beast).

One wonders if this evil spirit departed as a direct result of the music, thereby making it a legitimate medical therapy, or if God sent word to the spirit to cease and desist at least until such time as it is to be dispatched once again. So here we have two cases together with their diagnoses: the first medically advanced for that age, the second not so much, though this says nothing about the veracity of the latter's assessment (to which we can add "That's the spirit!").

THE SEVENTY PERCENT (RANGE) SOLUTION

We live in a world of amazing coincidences; optimal situations between too much and too little. Many of these occur in physics and have been logged before. They include the particle-anti-particle ratio, the strengths of the gravitational constant as well as the strong nuclear force.[1] Yet there are additional terrestrial balances not normally mentioned as often in the popular press. There are three I have in mind.

The first is that our own planet Earth's surface is roughly 70 to 71 percent water (and climbing). Is this a balance, however, that teeters on the brink of too much and not enough, excess and insufficiency? Well, consider the following. Ice ages come and go, dramatically lowering and raising sea levels, respectively. The Earth soldiers on in both cases. Though given the amount of time the world is in deep freeze, proportionately much greater than not, the default sea level appears to be low, which would not amount to our standard 70 percent. Thus the 70 percent mark might not be as delicate for the wellness of the planet as other factors. What would be and is becoming more severe is the temperature of the water itself and the effect this is having on rising sea levels and meteorological conditions.

A second example is the proportionality of nitrogen in the atmosphere, which lies at about 78 percent of the gaseous constituents. As is well known, the amount of oxygen in it, approximately 21 percent, with the remaining one percent being trace elements and compounds, would have detrimental effects were it to rise or fall. In cases where it might diminish, it would be difficult to get a fire going; were it to increase, we would have difficulty stopping fires from getting started, for then even damp wood will ignite.[2] Finally, our bodies are about 70 percent water. This is also delicate, for we can readily become dehydrated with injurious results, and we can also become too hydrated, leaving the heart and kidneys overtaxed with too much work to take on, thereby impeding the water from circulating and being eliminated efficiently. So it appears as though the seventy percent range is

---

1. Barbour, *When Science*, 57–58.
2. Lovelock, *Gaia*, 71.

optimal for at least three aspects of our living conditions here on Earth, the first not as impactful as the next two. More amazing coincidences.

Still another in addition to the 70 percent ones in view is the distance of the Moon from the Earth itself. The Moon now occupies just the right distance between the Sun and Earth so as to perfectly occlude the Sun during a solar eclipse. The reason for this is that, while four hundred Moons could stretch along the Sun's diameter, the Moon is four hundred times closer to the Earth than the Sun. The Moon's presence and position is critical for the Earth's biosphere. Not only does it affect the tides on Earth but also maintains the Earth's tilt on its axis, an angle of about 23.5 degrees, thereby giving us four seasons. But the Moon does not remain at its position, for it is receding from us at approximately the same rate at which our fingernails grow. Not appreciable from one year to the next, but will have a drastic effect on the order of billions of years, when the Moon will be released from the Earth's gravitational pull. The Earth will then wobble on its axis and climate will no longer be stable and conditions could even become uninhabitable. In the extreme, life on Earth, including ours, is closely tied to our astronomical dancing partner's distance from us. The question now arises as to how it is that we happily find ourselves on such a well-tuned planet, and cosmos for that matter.

Prior to embarking upon this question, there are other curiosities which science has isolated but has come no closer to unravelling. They are not coincidences as such but are no less amazing for it. Here are two of them. The first is the fact that for some unknown reason, the expansion of the universe began to accelerate roughly five billion years ago.[3] This normally requires an input of energy for it to occur. Where did that come from? Nor does anyone know why it commenced when it did, or that it did at all, nor do we know, let's face it, how long it will endure, for it admittedly changed once, so what would prevent it from changing again? But who or what gave it a push? The second is the anthropological occurrence of our pre-frontal cortex having become clicked on at about one hundred thousand years ago, at the time we became anatomically modern *Home sapiens*. We became thinking creatures. Not much, mind you, but thinking beings nevertheless. Thus an analogous question arises as for the first case, specifically, who or what flicked the switch?

A universe begins to accelerate its expansion, leaving its steady pace behind, and a brain becomes a mind and begins to evaluate itself and its environment, as well as ask probing questions that have occupied philosophers ever since (like how can I get more by doing less?). One wonders if the answer to both issues is the same, or at least similar. Cosmologically, in

---

3. Frieman et al., *"Dark,"* 12.

unguarded moments, we might be inclined to opine that the universe seems to be prepared for the emergence (or injection?) of life; the planet for the diversity of life; and primates for our arrival on the scene (known as the anthropic principle). Or are we simply reading back into universal history what we presume must have transpired in order for us to arrive here, in a deterministic or historicistic fashion—that the past could not have been different and hence eventually led to people, bright ideas, and credit cards?

Naturally, and I use the term strategically, the assumption on the part of many in the scientific community is that universes, planets, and organisms can accomplish these feats on their own, fostered by some dumb luck and being at the right place at the right time, without any outside assistance, and researchers will seek to corroborate this in their lines of inquiry. For some in the religious community, on the contrary, they have not closed the door to external involvement, or better still in their view, intervention. Even if this were to be the case, it appears at least to take some of the fun out of the search. But in the true spirit of the scientific quest, science is not about what we would rather have as the actual case.

EXIGENCY

According to standard dictionary definition, an exigency is something that requires immediate attention.[1] We are in the midst of one, and I am flabbergasted (pardon the lingo) that we have not taken steps to mitigate it. Given this, my remarks will be relatively succinct, for the debate has raged for decades and I can do little to contribute to it. I write of climate change, the bugaboo (again the jargon) of right wing movements, whether political or religious. We are being bombarded with propaganda from both sides, and both cannot be true.

What must be stated at the outset, however, is that the move of announcing that the vast majority of scientists hold to a certain viewpoint is what is known in philosophy as an appeal to authority, or in this case numbers or amount, for the majority can in fact be in error. Were this to be the lone card that is played in an argument, then by itself it would be insufficient. The emphasis should be on the quality of the research as opposed to the quantity.

I once spoke with a conservative fellow with whom I shared an interest in science. The discussion turned onto the polarizing issue of climate change. He, buoyed by the sermons delivered in his conservative enclave, argued that the topic is a hoax, owing to the fact that there are groups set to benefit financially from the ruse, namely solar panel, wind turbine, and geothermal companies for starters. This brings my blood to a boil. What in tarnation (all dictionary terms—feel free to look them up) do he and others like him think who stands to gain instead from the status quo? Not least of which are politicians who derive comfort from inaction (itself an action, or at least a choice), but mostly companies involved in the maintenance of fossil fuel consumption, those who clear cut forests and diminish biodiversity, together with converting forest into grazing land for livestock, specifically cattle, which, let's be honest, might as well be termed pre-burgers, to name but a few.

Detractors do not seem to recognize the difference between climate and weather, and that the former describes a multitude of the latter over extended historical periods. Climate, thereby, is many weathers strung together and averaged out. If I might be permitted a personal anecdote, I

---

1. Funk and Wagnalls, 223.

compose the following from the perspective of one advanced in years (respect your elders). On the positive side, if we may speak of it in those terms, in northern climes the turning of colors in tree leaves has been delayed of late, such that in my younger years, as I recall, the leaves began changing color in September and the show was over by Halloween. Which is why, when a friend and I were set to spend a day in a museum in town on November 11, then still a holiday in our homeland, I was surprised when there were still leaves that required raking and bagging. After a couple of generations, our summers and winters have grown milder. Nowadays it is typically not until the end of November when there is nary a leaf left on a tree, a one month differential over only a decades long time period (and what would decades be but a time period?). Great for those seeking to bask in the glorious autumn displays.

On the negative side, one might propose, this is indicative of an unhealthy planet, where temperatures on a global scale have increased, making for a greater risk of inclement meteorological conditions worldwide. Storms, for instance, of a greater intensity become more prevalent and inflict increased damage. We have been informed that in life we must take the bad along with the good, once a wise and globally applicable insight. Though, we must ask, even if we are contributing factors in producing the former, pervasive as, one could assert, it is developing? What we might now refer to as Indigenous summers, in an effort to be politically correct, are shifting, but the cost, the price to be paid, since it is virtually unavoidable, is that the advantage is overshadowed by its reaching prohibitive heights. The pleasure is not universal but instead is yielding a net detriment. Things have gotten out of whack and they need to be "put back into whack."

The difficulty in modifying minds and behavior about this concern is a political and economic one. There is no appetite or incentive to change if one benefits from business as usual. This is a conservative mindset. If something has not as yet adversely affected us or it occurs on the other side of the globe, then there emerges an inertia of will. Not even the arguments that our offspring will not only be impacted but imperiled by our current actions (and inaction) and will perceive our generation as having had the means but not the heart to make the necessary alterations to our method of planetary operation, that we were the last line of defense before the point of no return, that we have bequeathed to them a world where there could very well be dangers at every turn, what with the encroachment of disasters, that the thought had never occurred to us that our strategy of toughening them up because we had to struggle to achieve what we were able to build and amass ourselves and hence they should do the same is ill-advised in the face of the impending threats, that we thought so little of others and assessed their plight as self-imposed due to fecklessness on their part, that our borders should be closed to refugees of land devastation from drought and famine that cannot be traced back to us with less than two degrees of separation, since that would decrease community coffers and increase crime rates, how well do we expect this to reflect on us?

What report do we think later historians will submit on us, and, more gravely, how will we be judged? Shaking their fists at us will be the least of it. Evaluations of our indifference as mismanagement would be generous. Wretched more closely springs to mind. To have had the chance to repair what we have broken and to have turned our backs on it to the peril of others and subsequent generations is not merely irresponsible but vile. Vested interests are roadblocks to altering our course, for the mentality is such that there is too much at stake, meaning thereby our personal comfort level. Well, they are correct—there is much at stake, and more than we think or care to admit.

Should we not invest in the planet's welfare, all our other investments will be revealed as false hopes. And all of this despite the fact that, if we are honest, we are thinking more about ourselves rather than the planet and the effect climate change will have on us rather than it. Should the world be our main concern, then we need not be overanxious, for it has endured much worse and come out on the other side each time. This reveals that ecology and the environmental movement are anthropocentric, which they should be to an extent for the multitudes which will be affected by them. We do not want to be found as not having loved our neighbor, whether present or future ones. Yet the point remains that not everything is about us, but also about other creatures which might go extinct before their time. The Earth, on the other hand, has demonstrated that it is a survivor. The jury is still out on the human experiment.

Inveterate ideas are difficult to dislodge; once entrenched, beliefs seek only reinforcement, not counter-evidence. What will it take to revise attitudes? Personal loss for one. Yet we can act before it comes to that. The next generation is looking *to* us for help at this point; and if we act they will not look *at* us with scorn. I tend to become inspired to write when I get angry, so I must presently be severely perturbed. My hope is that ours will be the last generation to be afflicted with the malady of self-centeredness in relation to the task set before us.

My response to this conservative climate change denier was not to subject him to a barrage of climate change acceptor ammunition, because that effort would have been futile, armed as they usually are in apologetic-like fashion with prepared retorts of their own. But to make the move, as he did, of insisting on going where the data takes one, I seized upon that repartee to declare that he must then also be a believer in near-death experiences, a theme I have addressed elsewhere, since the data in favor of it are compelling and weighty, whereupon his facial features became distorted, he began to step away, and I could just imagine him thinking that I had gone off the deep end. Precisely what I wanted him to feel, as that amply describes my own reaction to his position. Sadly, he was unable to make the connection.

He disclosed himself as one who was decidedly unwilling to go where that particular set of data led. He therefore was not an equal opportunity data surveyor. In being selective about which data to be led by, he betrayed himself as unscientific. I did not even bother mentioning to him that the data are not exactly in favor of his classical theistic scheme either, also a topic I have treated elsewhere. Alas, we have not spoken since.

My travels have also taken me to the deep south of the U.S. where I engaged in vigorous debate with a fellow who holds that the pandemic is also a hoax. Why do I bother?, for the result is the same—futility. Part of our tendency is to look only toward those sources which bolster our own position. Despite the one-sidedness of their information, they are of the mistaken opinion that they have sufficiently done their own research, but neither blogs nor Twitter counts. I find it disconcerting that these deniers feel the need when their patience becomes exhausted to, in a degenerate manner, become reduced to shouting imprecations and sternly demanding that we bid a hasty retreat, though employing saltier terminology with a blue tinge. This becomes their reaction when they see it as their lone remaining move and thus puts an abrupt end to all further discussion. The exchange reminds me of Proverbs 9:7a, "Whoever corrects a mocker invites insult."

We differ in the content of our exigencies. The same fate befell this interlocutor as with the previous one. He sought refuge in his tribe and I in mine, where our views could suitably be massaged, thereby avoiding what both take as the uncouth element in the opposing side and instead gravitating toward those perceived as having an abundance of "couth." Due to the tribal tendencies on the part of many, there is no amount of information given to them that can prompt them either to alter let alone undo their position, and as such the stance becomes unfalsifiable and any attempt to criticize it remains wasted effort.

In a different instance, a woman among a group of protesters rallying against mandatory vaccinations can be heard announcing that she is against mandates in general. In an effort to demonstrate how this might unfold, consider the following hypothetical situation. Say it is to become mandated that at sporting events one is required to stand during national anthems, for up to this point it has simply been requested. What would her response be? Which trumps which, patriotism or individualism? If she were to abide by the mandate, then there is at least one with which she is willing to comply. If she does not and instead out of liberty elects to remain seated or, heaven forbid, take a knee, then she would find herself in the same camp as Colin Kaepernick (a robust alliteration). And it does no good simply to declare that standing was what she would have done anyway, since that was before the mandate; now it becomes an issue.

So if she is consistent, she will avoid standing and forfeit patriotism. If she makes an exception, she foregoes individualism and by her own actions reveals that not all mandates are negative. Best to avoid blanket statements, because they can force one into a bind, and those tend to be uncomfortable positions.

WHAT WILL WE THINK OF NEXT?

This segment is about science, though it might not seem so in the way it commences. Here is what I mean. Philosophers of science inform us that the mind is active in what we experience, such that it has a shaping influence on our senses even prior to encountering an object. This is because we operate according to our expectations—we believe the world will behave in a certain way and if and when it does not, we are surprised. As a result, we interpret the world before we experience it, to the effect that nothing is given anymore, there are just a collection of perspectives, none of which is objective.

Science has put an additional spin on this by alerting us to a few concepts of its own. The first which I intend to place in the spotlight is pareidolia. This is a common human feature in which we are pattern-seekers, which inspired the Rorschach ink blot psychological tests. Some, for example, have seen faces in the rocks on Earth, Mars, or on the Moon. In the mass of random shapes that, say, clouds can form, we have also been known to see patterns, like some taking on the form of, perhaps, certain Simpsons characters. Were we to point this out to others, they may see them as well, thereby confirming to us that we are not just given to vivid imagination but have similarity with the herd. That this is not rendered an objective image is disclosed when someone in the crowd announces that we are sorely mistaken and wonders how we can fail to manage viewing an obvious figure from the Peanuts comic strip. When we attempt to assert that this recognition is far from obvious, we might be met with a shrugging of the shoulders and reminded that we are entitled to our opinion, however wide of the mark it may be.

Biologists teach us that this penchant on our part must have had a selective advantage in our evolutionary history, for if we interpreted shapes in the forest as potential threats, ordinarily of the quadruped variety, then we would be in a position to tell the tale, assuming we were at the lingual stage, about the instance when it did turn out to be one of our natural predators and when we still had enough distance between us and it or them so as to bid a hasty retreat to safety. Those compatriots of ours who mistakenly declared that it was just our imagination were not regaling the community about anything anymore, leaving their gene pool decreased. Even in our current setting, horizontal images make more of a visual impression on us than do vertical. We can probably comment on how many cars there are owned by the people in our neighborhood but not as readily how many lamp posts there are. The reason is that in our distant evolutionary past, most of the threats we encountered were of the horizontal form ("lions and tigers and bears, oh my"). One can readily determine that this can be an adaptive trait for us, that those organisms bearing it left more offspring, and that the faculty is still with us today. I will have more to say about it, but wanted first to introduce the other concepts in view.

Another is apophenia, where those who have it are inclined to go beyond the pattern-seeking features of the former and maintain that there is more afoot than meets the eye. These folks insist that pattern-seeking has given way to something more insidious, more conspiratorial. Note that this is not, thankfully, a global human trait, for those persons have been assigned the label of conspiracy-theorists, once again, gratefully in the minority, who claim that certain groups or organization are combining their efforts so as to surreptitiously undermine what we might hold dear. Religious fanatics and political bodies are often cited as the culprits.

This is the only one of the concepts we are investigating that comes with a value-judgment, in that researchers state the target of our suspicions is not really there, that the identified perpetrators, in true Hippocratic fashion and contrary to the perception of those with apophenia, actually intend to do us no harm. In this way, the fear is little more than good old paranoia—there is in fact no one out to get us. Examples abound, such as governmental complicity in the events of 9–11, the military cover up of UFOs (now referred to as Unidentified Aerial Phenomena or UAPs), the involvement of the CIA in the assassination of JFK, the adverse health effects of the 5G network, and of course the genocidal purpose of vaccines (which would be analogous to parasitic organisms in biology. Parasites are more recent members of the biosphere and evolutionary scene in that they have not as yet developed to the stage of being in a symbiotic relationship with their host in order that both can benefit from each other's cooperation. For as it stands, parasitic activity often results in the death of its host, which is bad for business because it leaves the parasite without a host and at best it needs to find another or at worst brings about its own demise. Analogously, if vaccines result in death, then it would be bad for business for pharmaceutical companies to cancel out their customer base, for the dead are no longer taking or being administered their products. This would be a counter-productive business model and seems ludicrous to contemplate or entertain). There are also still some who believe that fluoridating the water supply is a communist plot. Those on each side of the fence cannot fathom how the other side fails to comprehend what the real situation amounts to and that it would be patently plain for any who care to have a look. The fact that there are at least two sides to this debate reveals that the actual circumstance is not so objective.

The final concept is not like the others. It is synesthesia, wherein some have the ability to experience in ways most of us are not accustomed. We ordinarily allocate our five senses into neat and tidy categories, but there are some who do not "see" it this way. Instead, they might hear sights and see sounds. I hesitate to call this a talent, but it can result, for instance, from someone having sustained a concussion or other head trauma. What results is an enhanced ability to experience the world, perhaps even to the envy of those not bearing the capacity.

Neither is it the case that those possessing it or them are any closer to an objective view of the world, for most of us would adhere to the notion that senses do not normally come in multiple versions, implying that synesthesia takes us even further into subjectivity, since their ability is not widely shared, though admittedly that response might be symptomatic of an anti-multi-sensory bias, and one does not in these times want to be caught committing the infraction of being an anti-sensory-pluralist. It can certainly benefit those with challenges in some way, for those who are blind can develop the skill of detecting objects in their vicinity with a heightened awareness of perhaps radar-like perception, our response being that we can manage quite nicely thank you very much with our, should we have them, fully-functioning five ("see" the alliteration?). Others, through being struck by lightning, for example, and having survived have sometimes awoken to the ability to play a musical instrument never previously attempted. The complexity of humans and their brains astounds.

Having introduced the concepts, I now wish to comment on them in the order of their appearance here. To begin with, one might regard pareidolia as having been necessary in the past and evolution as simply not yet having caught up to our contemporary needs. Yet there are dangers which still lurk in our current setting awaiting the opportunity to set upon the unsuspecting. These shapes do not always take on the form of quadrupeds, though this can still be the case when Dobermans are off their leashes, but can be of the bipedal variety and even of the two-, three-, four-, or in some cases eighteen wheelers. Except for those whose attention is distracted by I-phones and such, we could all benefit from seeing shapes in shadows even if we determine that they are made merely by trees, or imagining a skunk when it is just an outcropping, occurrences such as these often surfacing at night.

There is a difference, admittedly, between allowing one's downtime thoughts to wander and dwell on the benign shapes of clouds as we lie on the grass in a meadow, which itself is a distraction preventing us from something perhaps more urgent rising to consciousness, such as the increasing preponderance of ticks in northern climes, and venturing into the forest with only a club or spear for protection. There are, after all, laws and law enforcement for our at least limited assistance these days. And should they not be in our immediate environs, best to have 9-1-1 on speed dial.

But the point I wanted to make is that we are employing a present trait and imposing its salutary nature onto our distant forbears as something they must have been in possession of and which contributed to their longevity, ultimately yielding our arrival on the scene. We have them, so the argument runs, ergo they must have had them because they and we display an unbroken line of survival. This is a feature of the sub-discipline known as evolutionary psychology. The truth, however, is that we do not know what was in the minds of our primitive folk and how much they relied on these quirks, besides the more pressing drives of actually capturing prey. By the same token, the ability to perceive movement would also be adaptive, unless the figure of your interest is moving toward you, for by then it could be too late. The psychological aspect of these ideas is a theme to which I intend to return.

As for apophenia, if this can be directed, maybe then as now, toward an intuition as to the precursors for when a neighboring tribe may take to arms, then this has a decided advantage. For the amount of times somebody from one's community cries wolf and is ignored, though is correct on that very occasion when we have let our guard down, then it could be curtains for us. Best to take at least with a measure of seriousness those instances in which someone exclaims that "the British are coming," for you sure do not want to end up ingesting the ink from all that fish and chips wrapped in newsprint.

The points to be made from the foregoing include, for one, the fact that conspiracy theories can be true. We might not become cognizant for lengthy periods what the privileged know about certain events until perhaps a seventy-five year time frame when classified documents become declassified and access to information policies come into play, by which time those who could say "I told you so" have since met their demise. Each side thinks it knows the truth, but were we privy to covert actions on the part of others, we might change our tune. I suppose I am campaigning for a modicum of healthy paranoia, that I would not put it past some nations, governments, and assorted clandestine groups not to have our best interests at heart. The trick is not to claim that the "Illuminati" or the "Bilderbergs" are behind every shady occurrence. Best to keep an open mind until all the votes are cast, though I am deliberately avoiding the particular conspiracy of tainted election practices by forces domestic and foreign.

But this is not all. The other point I wish to make is that the insistence that humans naturally look for patterns is itself a pattern. There is a pattern on the part of some researchers to relate that human behavior must itself conform to some type of pattern so that we can make sense of how we got here. A grid has been placed on investigations such that the insights of evolutionary psychology must themselves yield patterns that cohere and conform to accepted (sub-) disciplinary canons. In essence, we have a tendency to look for tendencies. What would the new term be for the insistence on the part of humans that behaviors by previous humans must themselves yield a pattern, whether the claim that our imaging patterns must also have been a preoccupation on the part of ancestors and the scheme that others, again whether near or distant in time, have been schemers, hoping to expand their sphere of influence to threateningly include those dear to one's own respective constituencies?

Since many highbrow words have Greek, Latin, or German origin, allow me to coin one from the German, the one of the three with which I am most familiar—Vernunftgespenst. Like many of these terms, it is neither melodic nor rolls off the tongue. This is the price of accuracy, since the mathematical formula (or one of Murphy's laws) is beauty times precision equals a constant, meaning the two are indirectly proportional to each other, that is, the more you have of one, the less you have of another (anticipating a quantum point we shall submit later). Precision comes at the price of beauty and vice versa.

The word can be parsed, just like the conventional noun Zeitgeist: *Zeit* is German for time and *Geist* is German for spirit, hence Zeitgeist means spirit of the time or age. Similarly, *Vernunft* is German for reason and *Gespenst* is German for ghost, thus there resides, in my estimation, a specter in our rational faculty which teases us to be, coining another term, meta-patternist (I am under no illusion that these terms will catch on). Hence we find patterns because we expect to find them (also to be alluded to when covering quantum), and our expectations are usually met, whether they are actual or not. We might believe, for instance, that some persons are hostile toward us, and with this perspective, their speech and actions cannot help but contribute to and confirm our suspicions, again, whether legitimate or not. Should these characters be in league with others, we have before us a conspiracy. We hypothesize that our anthropological forebears operated this way too, though establishing that is problematic. Nevertheless, we are undeterred. It is easier to live with a pattern we are comfortable with than to deal with the messy or chaotic. We, like the God of the Genesis text, are order-imposing beings. These sub-disciplines attest to it.

It should be noted that apophenia and synesthesia are accepted categories of biological and medical practice as well as conditions some humans face, but what is regarded common to humanity is pareidolia. I am suggesting that apophenia might also be placed in the category of the common, for if we cannot be definitive about what our primitive relatives thought, however likely it might be, neither can we then conclude that they did not think in conspiratorial terms, thereby making apophenia not a pathology but also a tendency. We simply cannot tell and ought to be forthright enough to admit as much. That would be a pattern and tendency I could endorse.

One more thing before we close this section. There is another potential ingredient to the recipe. What is also a common human feature is something psychology has once again taught us, and it is known as a Gestalt switch, where *Gestalt* is German for figure, form or shape. The gist of it is that when we are shown an image and are asked what we see, most of us initially see only a jumble of light and dark areas and cannot make anything recognizable out of it. When we are then shown what lies in the jumble, we wonder why we could not make it out before, and we will never fail to see it again. It was once pointed out to me by an academic dean of a Christian college that a wood carving on his desk contained a hidden name—that of Jesus—and mostly Christians were able to make it out. That was one test of Christian membership for him. Not a reliable weathervane in my view.

Sometimes there are two images contained in one jumble: in one case, a chalice appears and on the opposing sides appear faces in profile. In another case, one is of a young girl and within the same image lies an old woman, depending on how one looks at it. Upon recognition of the two, we are then able to switch back and forth from one to the other with ease.

The question which needs to be asked at this point, in tune with the explanations given by researchers, is has this ability assisted us in our evolutionary history, or is it merely a product of what we can reproduce with contemporary graphics? Was there a primitive counterpart to this? Once again, should we be able to see in the clouds the shape of a Simpsons character and then click on to a Peanuts figure within the same form, then this would be an instance of pareidolia, but with a twist: now it is of the type where we can see multiple images. Switching back and forth in a Gestalt manner from one kind of image to another is not necessarily adaptive if our concentration is distracted from a real danger, were we with amusement to focus on the wrong image at the wrong time. This would not support the evolutionary psychology thesis and hence does not provide us with a thoroughgoing selective advantage.

According to their view, it helped us survive such that evolutionary psychologists could arise so as to inform us of it. But this would only reinforce that portion of the population which switched correctly. Yet here we are, switching back and forth, all the while being distant offspring of the correct switchers. We are kin to those who survived, for incorrect switchers died out. Perhaps we have successful gamblers in our ancestry. On this logic, it explains the emergence of numerous casinos. My point is that this subdiscipline can be employed to support one viewpoint as successfully as it can another. We are present and display certain behaviors, therefore we must have exhibited them in the distant past. This is a self-fulfilling prophecy.

The steps in their (il)logical scheme can be listed as follows:

1. humans perceive, think, and act in specific ways;
2. this has been the case for as long as we have recorded history and even before;
3. all of this persists to this day;
4. we have arrived at this point because natural selection, as its name implies, has selected for the advantageous;
5. we and our traits would not be here were they not adaptive;
6. therefore our distant forebears must have behaved this way as well in an unbroken line of survival and offspring all bearing these traits;
7. therefore these traits must have been adaptive once they appeared on the scene, for those bearing them must have outcompeted those without them;
8. therefore this explains in turn how we came by way of them, but we are not as dependent upon them as we were initially, so we can consider some of them vestigial.

Difficulties with the argument include these elements:

a. ever since the onset of agriculture, non-nomadic and hence sedentary lifestyle (by this not intending the origin of flabbiness on our part or even the lack of cardio-vascular activity), and civilization, roughly eleven thousand years ago, the above traits have undergone reduced import;
b. this time frame, while insufficient for a significant amount of evolution to occur, can have some effect (lactose tolerance occurred within the last five thousand years), and of course some wolves, for example, have become transformed eventually into about four hundred breeds of dogs—the earliest evidence thus far being traced back to Belgium at least thirty-five thousand years ago;[1]

---

1. Stringer, *Lone*, 166.

c. the continuity between what was significant for our distant ancestors and is less significant for us now could very well have met up with mutational variants not including, or including fewer of, these traits as time went on;

d. since the two would then be in less direct competition, we could very well observe individuals not bearing these traits;

e. we do not, though there might be varying degrees of them, then as now;

f. this would be an additional instance of current selective neutrality;

g. why is it that when we see images in the clouds it is hardly ever those things that can harm us, whether human or other animal? (though malevolent shapes are in fact detected there at times);

h. should we not be seeing more of these had they survival value?;

i. is it because fluffiness does not provide the right canvas for that which is threatening?;

j. might we not be more wary of sounds than sights, like things that go bump in the night?;

k. these days assailants, such as snipers, strike from a distance, so no amount of seeing forms would assist us there;

l. this can also be extended back in time to the dawning of the age of launching spears and shooting arrows, again doing their damage sometimes unobserved from a distance;

m. for all the things that could have harmed us then, why were all of them not seen in some shapes somewhere, and why is the viewing selective, that is, not a full range of threats?; and finally,

n. if seeing images in natural things is, well, a natural thing for us, then why do many of us have difficulty finding a particular figure, such as Waldo, in a picture among the welter of other shapes (excusing for now the fact that these images are not natural but contrived)?

Thus far, the difficulties outnumber, though not necessarily outweigh, the steps in the argument by a factor of nearly two. In my appraisal, evolutionary psychologists seem to expect that they can explain more than they are able. Theirs is not a logical requirement; the actual history may have been markedly different and evidence either way is sadly lacking. First moral of the story: best to think through what might be unduly convenient theories in case what is disclosed may become inconvenient. Second moral: we need to be vigilant in placing stock where investment is not warranted. All I can counsel is *caveat emptor*—let the buyer beware.

As an addendum, and permit me to begin by relating to the reader that I simply cannot help myself—I was once told by a professor in graduate school that those who are paid to talk often work overtime, and it turns out he was right as I see the same in my own career—when analyzing body language, a person's gait, and facial expression, which admittedly are hard-wired into us from the early stages of our species, we now have a more scientifically informed way of reading what others are feeling and perhaps even intending, though the investigation is still in its infancy. Even our efforts to mask these traits fail to resist their rising to the surface and manifesting in facial features—we will ultimately betray our emotions despite the attempt to conceal them, due largely to this hard-wiring. There are certain muscles in our faces, for example, which tell a story we cannot hide, unless we are highly trained. These behaviors transcend time, space, and culture, yet contrary to these studies, there are those who *are* highly trained and they are known as actors. Evidently, these emotions can be hidden as well as duplicated, which is how actors earn their pay. The best of them are effective at displaying emotions that are not specifically their own but the persona they portray, and we suspend our judgment and go along with the ruse, at least for the duration of the program.

This is the difficulty with interpreting intent, for we have no alternative or recourse but to gauge internal states through external cues. This is comparable to officials in a football game peeling away a mass of humanity in the attempt to determine at the bottom of a pile which team has recovered a fumble, the football being analogous to one's intent, with the exception that an assessment can be approximated in this sport but not so readily in the aforementioned. Even were persons to report on their interiority, the accounts might be inaccurate at best and deceptive at worst. Though here we are, especially in courts of law, ruling on intent and issuing sentences based on it. Despite its being subsumed under the umbrella of science, it is not an exact one. What precisely is in the heart of those accused of a crime? Can their testimony be relied upon? Can they be trusted as contributing members of society? Or do they kidnap and hold captive people in their basement as did the former marine in Cleveland? Neighbors were surprised to learn of his crimes, suspecting that he was nothing other than an upstanding citizen and participant in the community. How could we have gotten it so wrong? Well, because some people are adept at hiding their true selves.

Perhaps evolutionary psychologists sometimes are not looking in the correct places and it turns out that the skill of covering up our actual selves is the true survival value. Even Hitler was charming to his international guests, who would later become his enemies, and he was to engage in murderous acts. And British Prime Minister Chamberlain thought Hitler was on the level and was taken in with the signing of the non-aggression pact. Hence, playing the part of devil's advocate, this could also be argued for as the cause for the survival of our species. Admittedly, the more likely scenario is cooperation, not just competition, as that which accounts for our societal longevity (which to date in evolutionary terms is not extensive). Yet human history has also been peppered with might making right. Perhaps Machiavelli was correct at least in part. The point to be made, once again, is that the sub-discipline under our microscope could benefit from some refinement.

(One more thing, where did we obtain our sense of taste when it comes to color schemes concerning which colors go with which and which others do not? They appear to be pervasive in that most persons agree with the assessment. But are they innate or learned?)

VISITATION OR DESPERATION?

I enjoy watching the History Channel as well as H2 (which does not stand for one of the two constituents of water) for the documentaries they broadcast. But so as not to seem too highbrow and to appeal to as wide an audience as possible, it also includes in its programing lineup shows such as *Pawn Stars* and *American Pickers*. It further contains offerings on the enigmatic side of life, like the *Curse of Oak Island* and others devoted to unsolved mysteries as the *UnXplained* and *History's Greatest Mysteries*. There is one telecast, though, that could be judged as out of place, or, as *Sesame Street* (not one of History's series) would intone, "one of these things is not like the others." I write of *Ancient Aliens*.

It is an interesting show in that it provides the viewer with contemporary science and even interviews with world renowned physicists such as Michio Kaku—one of the founders of string theory. For roughly the first half of each episode, one is brought up to date on matters of scientific interest and its recent developments. It then abruptly changes gear and speaks of possible accounts of occurrences which are quite literally out of this world. The broadcast takes the daring step of diagnosing the enigma in view as explicable only by making reference to ancient astronauts who have, since their arrival on earth in the distant past, influenced the course which historical events on earth have taken—a type of extra-terrestrial intervention.

Stopping short of declaring that God and angels have had a hand in history, the interviewees conclude that these ancient aliens were described as "the gods" in the texts left behind by ancient civilizations. In fact (and I employ the term loosely) their visitation so impressed the ancient mindset that these events have been committed to writing, and the intruders were accordingly venerated in architecture, artifacts, and hieroglyphics to commemorate their contribution and with the expectation that they would return. Regrettably, however, I have yet to catch an episode on any of these topics which is definitive; the audience is always left hanging not with fact but conjecture and hoping that the next episode will be conclusive. This is how they keep their audience watching.

Having already broached the topic of our pre-frontal cortex being clicked on at a certain point in our distant ancestral history, and firing another volley into the potential ET skirmish, what was it from about 10,000 BCE onwards, when the ice sheets of the last ice age were retreating to the poles, that ignited (or thawed) in humans catapulting them into an architectural frenzy and erecting temples and other structures ever since, most of which seemingly requiring no ordinary ability to construct, and hardly attempted prior to the last deep freeze? Why then and not before? And should we happen to like neither coincidences nor conspiracies, what recourse is left open? Are we then confronted with the pressing need to play the ET card?

Speaking of architecture, some of the world's great wonders, as the long-standing pyramids of Egypt, notably on the Giza plateau, are thought to be not only intended as memorials to eminent humans such as the Pharaohs, in this case Khufu in the largest of them (more recently determined as faulty since few Pharaohs are located in them, quite possibly owing to the looting of grave robbers, rather most are located in the Valley of the Kings in Thebes (present-day Luxor)), but also monuments to the gods, sometimes pointing to their origin in the cosmos, as a kind of beacon should they ever decide to make another appearance (as if they required a type of interstellar GPS system), but also to stand as an eye-popping example of how humans could not have accomplished the feat without their assistance. They endure, ancient astronaut theorists urge, as a reminder of the past glories of alien interaction with humans, specifically bestowing the technological wherewithal to complete what allegedly might not otherwise even be attempted in the contemporary scene.

Taking as still another instance is the perplexing enigma of Stonehenge and other stone structures like them found in many parts of the globe, for with them we perhaps have before us feats of engineering we can scarcely duplicate even today. We can only marvel at their construction and are left asking how Stone Age peoples, assuming the dating is correct, were able to achieve them. Aside from the knee-jerk reactions on the part of ancient astronaut theorists, claiming as they do that those civilizations were the recipients of extra-terrestrial support, some researchers in the present day toy with possible solutions. Concentrating on Stonehenge for the moment, the largest stones stem from the lone location in the U.K. where they could have originated, namely a quarry in Wales roughly one hundred and forty miles to the West of where they were transported to the Salisbury plain. One ingenious proposal was the use of felled tree trunks near the quarry site as rollers. The idea is to push and/or pull the stones on top of the rollers and once a trunk is no longer under a stone it is then placed in front where it can dutifully resume its rolling obligation.

Upon contemplating this hypothesis, detractors assess that given the sheer weight of the quarried stones, reaching into the many tons, they would crush the would-be rolled logs. Both undeterred and undismayed, other researchers attempt to repair the breach by hitting on the notion of employing the wood in an alternative fashion. Instead of using the tree trunks as rollers, they would be cut into shape with stone tools, which the researchers confirmed by doing the same, thereby leaving (pardon the pun) no stone unturned. These would then be placed lengthwise under the edges of each stone and fastened into place with cross planks, themselves held together by boring holes into them and "nailing" them with wooden pegs and employing cords of rope for sturdiness, thus preparing them for pushing and/or pulling them as one would a sleigh. Subsequent to one hundred and forty miles of huffing and puffing on, dare we say it, bad road, the stones would be delivered to the intended spot, at which time the researchers would wipe their hands of the affair and declare the problem solved.

Yet the difficulties persist, not least of which is how these timbers would escape the crushing that those in the former instance could not? They neglected to explain this detail. They did not propose how the use of fewer logs could withstand the even lesser weight distribution of the stones and not more readily crush the timbers. Others include how holes could be bored in the absence of drilling, which would be anachronistic, in a way that would not split the wood and with which the pegs could be snugly inserted? The researchers did not venture this information either. Not to mention (why do people say this when they have every intention of mentioning something?) how the largest delivered stones would then be set upright and, most mystifyingly, having a third stone placed on top of some of the two, thereby making each set of three resemble the Greek letter pi (incidentally, when I am asked what I wish for dessert, I sometimes respond with 3.14159 . . . , even though it multiplies the syllabic effort by a ratio of 7:1, my rationale being the striving for effect more so than economy of verbal expenditure. Regrettably, the strategy does not always have the desired result. Yet another perplexing puzzle.)

Hence the problems remain unresolved, meaning the ancients were no slouches and had greater abilities than we give them credit for, revealing them as the actual ingenious ones and having one-upped us in the enduring battle of one-up-manship (-personship?). But while we are on the topic, a further suggestion to explain how the stones were transported and lifted into place involves the science of acoustics. The way it supposedly works is through a range of sound frequencies which have a distinct anti-gravitational effect. Phonons, the sound equivalent of the photons of light, can actually travel upwards contrary to gravity and levitate objects, though to date the phenomenon affects merely lightweight objects such as ping-pong balls. It must be admitted, however, that in theory the levitation need not be confined to the lightweight. Sound can already move grains of sand into artistic wave patterns, so what would be the limit of its powers and could the ancients have tapped into it where we, at least as yet, have not? This leaves us with the question as to whether phenomena like these can be counted as within their skill set; if so, then what else were they capable of?

There are those, therefore, who would assess the idea of alien direction of the course of human history as hasty. The program takes the viewer on a ride which culminates in the exasperated announcement that there is no other recourse available to us as to decode how world history unfolded, meaning we are compelled to making the lone move of acquiescing to their proposal, for there is no other option. We should therefore be courageous enough to accept it. I remain, however, in the unconvinced category. Take heart, dear reader, for the deck could very well contain other cards. As mentioned, the difficulty which surfaces is that sometimes such architectural feats are not only unequalled in times past but might well-nigh be beyond our reach even today. Despite our contemporary sophistication, we would be at pains, hard-pressed, or even incapable of duplicating the prowess which the ancients displayed, enhanced as they apparently were by the tutelage of those astronauts. This plays right into the hands of the show's interviewees, for our failure to bring forward an explanation as to how they achieved such high caliber engineering feats empowers them to proclaim that these structures had a greater-than-human origin. This approach, however, suffers from improper reasoning in at least two senses.

The first pertains to the coordinates of terrestrial location. The whereabouts of the Great Pyramid at Giza, it is asserted, is mathematically significant, lying as its pinnacle does at the latitude of 29.97628 degrees north. The figure, by a startling coincidence, agrees to at least ten decimal places with the speed of light when measured in metres per second and by the appropriate order of magnitude. The problem with this measurement, though, is that it is arbitrary. To begin with, the length of a metre is not a natural law but has been agreed upon by convention. And in terms of latitude, the earth's equator is non-arbitrary, for the middle remains the middle, regardless of the movement of the tectonic plates, and there will always be a middle. Beyond this, however, there is no law of physics requiring the value at the poles to sit at 90 degrees other than the customary one of right angles describing that amount.

But this need not be the case, for we could conceivably opt for a metric scheme and designate the poles as 100 degrees were there to be consensus on the issue. Should we elect this alternative, the figures would not be as impressive. Nor, presumably, did the Egyptians have these specifications in mind when conceiving of the edifice. Even more so with lines of longitude.

Pointing once again to convention, a circle, or in this case a sphere, bears a circumference of 360 degrees, which also need not be the case and could be assigned a different value. But most importantly, there is also no law of the Medes and Persians dictating that line zero degrees go through Greenwich, England, as those in Britain being the first to contemplate it naturally chose their own nation as deservedly occupying it. Once again this amounts to arbitrariness, nothing demanding what has been internationally agreed upon. As a further example, consider the idea that by looking west from the westernmost island in the Alaskan Aleutians to the easternmost Russian island constitutes viewing tomorrow is also entirely arbitrary, for it is completely dependent upon the International Date Line running between the two, something that *we* assign to the globe.

What is more telling is the other instance of faulty reasoning. Ancient astronaut theorists leave us at a loss at evaluating the glaring mysteries that surround us with an explanation other than ET has come calling. This is what is known in informal philosophical logic as an appeal to ignorance. Should there be no other account on the horizon, then we must throw our hands up in despair, they would insist, and, confronted by the seeming futility of the affair, summarily posit ET as the cause. This commits the fallacy of adjudicating that if all else fails, then by a process of elimination, whatever remains, however improbable, must be the correct one, to which we can retort, "not so fast, Sherlock!"

To the emphatic question "what other recourse can there be?" perhaps reveals a lack of imagination on our part together with a failure to confer on the ancient peoples a level of ingenuity we have not as yet uncovered, for we ought not to be too quick to conclude that there were not brilliant minds to have emerged then as now. Else we could be found to hold and be charged with a "recentist" prejudice when it comes to our ancestors, believing we have exhausted all the possible explanations and nursing a perceived ancient inability. The resolution could very well be that they were smarter than initially suspected, so give them some credit. As the old saying goes, they had a need and this facilitated invention; we do not have such a need, so invention has eluded us (not even mothers can help us here). The inability to fathom how they managed it would then appear to be ours. We might be too hasty to abandon the search for more mundane answers, one example of which is the Coral Castle in Florida.

In an area south of Miami in the earlier part of the twentieth century, one lone man built an entire complex of structures from the coral geology of the ground on which the complex was erected. No one to this day can fathom how it could have been accomplished, an instance of which is a many-ton swiveling door to the complex which can easily be opened even by a child. This is a single individual working secretly at night using, by our standards, relatively simple tools and scaffolding, marking an achievement we know not how. At least it demonstrates that the seemingly impossible can in fact be carried out to completion by the usual suspects using customary materials, and not all enigmas are ancient. Three cheers for doing extraordinary things with the ordinary. Now the inclination on the part of ancient astronaut theorists, of course, is to respond that this builder was informed by the secrets of ancient wisdom, which itself was imparted to them by intra-galactic travellers. Sometimes you just can't win.

At the same time, and in all honesty, however, in an effort to mollify the critical statements being made here, I must object to the conservative mentality on the part of some scientists in the history of the study of our universe. In the 1960s, when certain astronomers/astrophysicists dared to speculate about the existence of planets and solar systems similar to our own (the famous Frank Drake equation outlining the probability of this occurrence had been devised in 1961)[1] and thus promoting the possibility, in addition, of life elsewhere than on Earth, these unfortunate souls were ridiculed for their foresight and scolded for believing in "little green men."

Nowadays, the notion that "we are not alone"—the title of a work by journalist and science editor of *The New York Times* Walter Sullivan from 1964 (subtitled *The Search for Intelligent Life on Other Worlds*), who attempted to undo this injustice—is the accepted mindset, where there is a vigorous surge to be the first to locate such an exoplanet, with the understanding that they could very well number in the billions. And all of this activity without so much as an apology to the trailblazers and how they earned censure at the hands, and mouths, of the former naysayers. Conservative attitudes can tend to become recognized as short-sighted with time. Will something analogous turn out to be the case for extra-terrestrials, where the current scorn heaped upon believers includes referring to the beings not as green but rather as "grays?" For now, admittedly, there is a vast difference between locating an exoplanet and actual contact, but the mysteries nevertheless linger.

As an aside, what I can state in their favor is my aversion to the notion of panspermia, initiated by Fred Hoyle—the idea that Earth was seeded by rock ejecta having been thrown off the Martian surface from collisions it experienced, and eventually raining down upon Earth, leaving biomacromolecules and/or organisms to proliferate here. Another version speaks of spores being carried on asteroids or comets.[2] My disquiet toward this view

1. Sullivan, *We Are*, 246.
2. Randall, *Dark*, 221–22.

is that the materials which showered Earth would either have been burned up in its atmosphere or obliterated once they collided with it. DNA, for instance, is readily denatured by processes already here on Earth, thus what chance does it have, risking an appeal to ignorance myself, of remaining intact when arriving from elsewhere? As well as how it would get underway in a functional manner once on the scene, that is, not be inert.

Admittedly, this is not the case with amino acids—constituents of both DNA and RNA, which "can survive comet impacts or be created when extraterrestrial material hits the ground." (The issue then, however, would be whether the amino acids are biologically active, for whereas the ones on Earth bear a chirality or handedness of being left-handed, those on space rocks contain both right and left varieties).[3] Plus the origin of life question would merely be placed one step further back by requiring an explanation as to how it got its start, then, on Mars. Despite my allergic reaction toward the ancient astronaut theory, remarkably, I think it has more in its favor than what is known as the directed panspermia version of the theory, championed by Francis Crick, where the complexity of the DNA molecule reveals, in his estimation at least, some type of design, since he does not consider the length of time life has been on Earth to be sufficient to assemble all of DNA's complexity. Later ancient astronaut theorists proposed who those designers might be, namely extra-terrestrials implementing their design on Earth by seeding it themselves. The unlikeliness of these scenarios is the degree to which I object to the panspermia versions as they currently stand.

But there is more that needs to be stated. What the theory of panspermia has going for it includes the following. In the early ages of our solar system, the habitable zone—the region where the optimum conditions reside for the onset and proliferation of life—stretched from the Earth to Mars, making two planets as potential homes for life. Since that time, the zone has shifted inwards toward the Sun, leaving the Earth as the lone planet in the right place at the right time, known as the Goldilocks zone, for life to abound. Another factor is humans appear to work on a greater than twenty-four hour cycle—our biological clocks are geared closer to a twenty-five hour diurnal period, which, coincidentally, is the same as the rotational period of Mars; it turns on its axis completely in 24.6 hours while the Earth's is shorter by more than half an hour. This biorhythm has been confirmed when astronauts return to Earth from extended visits to the International Space Station (ISS).

Additionally, humans need to wear sunglasses to shade their eyes from the Sun, which would not be required further out on Mars. Similarly,

---

3. Randall, *Dark*, 224–25.

humans are in danger of becoming sunburned if out in the Sun unprotected for too long, once again not problematic at Mars' distance. Finally, humans seem to operate best physiologically at an external temperature of sixty-seven degrees Fahrenheit, where the healthy forms of fats are produced, whereas at higher temperatures not so much. As one might anticipate, the temperature at Mars' equator in summer is this very one.

Convinced yet? Well here is what is not going for it. First, at least in part, for the amount of time human ancestors have existed on Earth, this would have provided ample time for evolution to work so as to modify these early primates from one rotational period to another, were it to have provided a more adaptive biological clock, though it would have occurred in the opposite direction for Earth, in that its rotation slowed, not hastened. Other organisms would need to be studied in order to disclose whether the same holds for them. Hence the retention of a twenty-five hour cycle perhaps is selectively neutral and by itself fails to confirm that our previous address was on Mars. Second, over the course of time, our Sun has become brighter and hotter by double-digit percentages. Natural selection has not as yet caught up to this increase, so the technical miracle of Polaroid lenses will have to do, and they do so nicely. Plus a third, we are more prone to sunburn not because Mars was our initial home, but since we have lost most of our hair from our more simian days. Once again, technology in the form of chemical sunblock (as well as stylish hats) has come to the rescue, yet so as to beat the heat, better to find some shade.

Lastly, we can produce healthy living temperatures if we wish through the technology of climate control in our dwellings and other meeting places. There are, however, a greater number of temperate areas on Earth than Mars could supply, the issue being not merely temperature but the other harsh conditions found on Mars, such as what little atmosphere exists is carbon dioxide, the exposure to the Sun's harmful rays as a result (of the thin atmosphere, not the CO_2), and the dust storms that pummel the surface at a torrid pace. The Martian atmosphere, as it turns out, began to bleed into outer space since Mars did not have a magnetic field to keep it stable, for it lost its interior heat as did our Moon. (The gravitational pull of the Moon on the Earth is what causes the molten magma in the Earth's interior to slosh against the mantle, the friction creates an electric charge (there are also electric currents under Earth's surface), which in turn forms the magnetic field and its poles. These poles function to keep harmful solar and cosmic rays from reaching the Earth's surface and causing damage. Both fields and poles shift periodically and sometimes even reverse.) Additionally, solar winds further impacted the surface, leaving it a desolate place. Why then do we imagine we could make it any different by terraforming it? This would neither stop nor reverse the bleeding, since oxygen gas is lighter than carbon dioxide and would not be captured by the planet or its atmosphere.

Panspermia-backers need to be careful here not to mix the current with the previous conditions on Mars. If optimal temperatures for humans obtain on Earth, then it does no good to declare that this particular condition can also be found at a narrow band on Mars now when its past is the major concern, for that is when we hypothetically resided there. My recommendation for them is to be consistent as to whether they are making an appeal to Martian conditions now or then, for it would augment their argument. The distinction is crucial, for the argumentative thrust pivots on the timing chosen for Martian evidence. At some points, the distant or remote in time Martian scene is invoked; at others the more recent. This about face, even vacillation, weakens the force of an otherwise interesting if not compelling case.

Besides, place yourself for a moment, for argument or debating purposes, in the shoes (Boots? Sneakers?) of purported ETs. Would you impart your far-advanced technological knowledge to humans who have not as yet demonstrated that they can be trusted with their own primitive by comparison military might, and where Nobel Peace Prize laureates *for peace* turn around and persecute a religious minority or order an incessant amount of drone strikes on suspected terrorist holdouts, together with the already outlined (non-Nobel) brutality toward ostracized peoples by domestic law enforcement? If I were to be the one issuing the ET commands, I would regard the human population as on probation and defer the transfer of any sophisticated information until such time as these creatures can prove themselves reliable and trustworthy with it. Perhaps this is why they have not come out of the closet for untold millennia. I wouldn't either. We have not as yet earned the privilege.

As a final instalment on astronomical themes, we invite on to the stage the discipline of biology. Here is why. Throughout most of the history of Darwinian thought, it was assumed that the directionality of organismic change occurs from genotypic modifications in DNA through mutations, stemming from radiation, chemicals, and just the plain old sheer luck of errors in replication, to phenotypic expressions of them in the structure of the organism, hopefully making it more attuned to its environment. The very notion that the process could occur in the reverse direction, from environment to form to DNA, was deemed heretical, doctrinaire lot that evolutionists can be. I continue to be struck with the dogmatism of Darwinism (note that an "ism" is an ideology, not a science)—that Darwin should be treated as infallible and should be defended at all costs against the unwashed, by whom they mean creationists, for these are the only two alternatives: if one is not for Darwin, then one is against him. Creationist fanatics, after all, in their view would be the only ones who could be the detractors. All the while, physicists, on the contrary, have no trouble pointing out, say, where one of their heroes—Albert Einstein—was in error. So lighten up, for it's easier to learn something with an open mind.

In any event, it seems as though the heresy has something in its favor: it happened! An American astronaut who returned to Earth, landing in the desert of Kazakhstan, spent a record five days short of one year at the ISS. Upon his return, he was examined to determine how this period of weightlessness affected his anatomy, physiology, and biochemistry. Researchers found that his DNA had actually altered by an astonishing seven percent, affecting his immunological system, among other things. An even more protracted stay would also cause our legs to atrophy, for instance, unless we are careful to place upon them the requisite stressors so as to assist in their upkeep. Hence whereas it was thought that biological alterations occur from the inside out, internally to externally, it now becomes clear that this is one more area where we need to take a both-and as opposed to solely an either-or approach.

Finally, the anticipated theological question from the conservative contingent is the obvious one about whether the Messiah would then be required to attend to the fallen races spread throughout the universe and be sacrificed for the missteps of each and every exoplanet ever having arisen, though we are informed that this was to have happened but once (Heb 10:10) (Unless of course the aforementioned possible early view of Paul is correct that the Messiah began his existence as an angel, in which case he would have been just one messenger among many others on a salvific mission. But then there would be an equal number of them, should they all have been successful in their task, to have become exalted to God's right hand. Reassuringly, then, space must not be a limitation in the heavenly realms so as to accommodate all of these right-hand seated or standing beings (Ps 110:1; Matt 26:64; Mark 14:62; Luke 22:69; Mark 16:19; Acts 2:33; 5:31; 7:55–56; Col 3:1; Heb 1:3; 8:1; 10:12; 12:2; 1 Pet 3:22). Admittedly, the last reference in this list could potentially militate against the position if the term "other" could not legitimately be inserted immediately prior to "angels.") Otherwise it would make for a Messiah stretched too thinly were he to do all this heavy-lifting himself and carry the scars from each round of sacrificing. Yet this assumes that all races have succumbed to temptation and have gone astray, thereby calling for God's ultimate front-line worker. But wouldn't it be a kicker if we were the only ones who needed it! I hope that would not mean that we will be ostracized by these other races at the afterlife lunch tables.

PART 3

Religion and Science

QUANTUM CONUNDRUM AND OTHER RIDDLES OF SCIENCE AND RELIGION

Physics and Philosophy

Assumptions can get us into trouble; recall how parsing the term "assume" highlights the difficulty. But making them is part of our method of operation. We assume that the universe is sufficiently orderly so that we do not need to test the engineering and design that went into the production, say, of a chair every time before we feel confident that it will bear our weight and are not surprised that we can commit the welfare of our corporeality to being seated in it. Instead, we sit without thinking about it as we have many times before, faithful chair that it has been, unless, of course, wear and tear have diminished its capacity. Yet will it be loyal this time or the next? We seldom give that question a second thought. Assumptions, therefore, work until they do not (you may quote me). There will inevitably come a time when chairs will no longer meet their obligations, but this is not something about which we need carry anxiety around with us. Chairs are usually reliable, so we think we are pretty safe in our assumption, at least for now.

The history of science is often marked by the history of assumptions. As mentioned, assumptions are adopted until such time as they sputter, then cough, then backfire. At this time, confidence in them diminishes and we wander in search of a replacement. It is characteristically not until such time that we really become conscious of the fact that we have been operating with a set of them all along. We do not notice them until they come up for review. So where do we turn, now that our secure moorings have been removed? We are then forced into a position where we need to think about them seriously. Our comfort level has dropped. Should we feel betrayed?

There are some assumptions that are worth examining to determine whether they live up to their billing, since when it comes right down to it we cannot even remember a time when we actually cast our ballot lifting the current set into the ascendency. We thought it just came with the territory, the normal everyday course of events. Well that has now become upset. The world we were used to no longer seems to be worth investing in. Who will rescue us from this insecurity? Let's investigate certain sets of assumptions and determine how they relate to theological questions about the divinity. I purposely employed the terms determine and relate here, as they will have particular significance in this pursuit. And depending upon the philosophical and scientific background of the reader, the following will be either an introduction or a refresher.

Newton

We commence with Sir Isaac Newton, touted as one of if not the greatest scientist of his or any age. He set about to compile a list of assumptions that his contemporaries could invest in for operating in the world, nor did he envision a time when they would not apply. He, like us, sought a metaphor upon which his model of the universe could be hitched. We often resort to the use of analogies when no other descriptions will suffice. So give him a break. We tend to understand, say, electricity and the electrons which carry the charge as producing a current which then flows through a wire, though electrons do not flow; only fluids flow and electrons are not a fluid. This does not prevent us, however, from persisting with this handy analogy.

Isaac (let's call him Ike) inherited some ideas from the medieval outlook and incorporated them into his more modern one. They are listed below.

1. The world is experienced or perceived realistically; that is, basically, what you see is what you get. In essence, the world is accessible to our senses and the latter suffice in the ability to describe it.

2. The world is rational in that theories articulate the world as it is in itself. That is to say, the world is reasonable and intelligible, can be understood, and is coherent, for it makes sense and can be known objectively, meaning it does not differ from perceiver to perceiver and as such does not depend on anyone's perspective.

3. The world is static in that, despite its changes, as in the four seasons, there is no net change in it. Accordingly, the world displays limited motion. The following four (plus) points then constitute Ike's own contributions.

4. Space and time are held to be frameworks where every event is located objectively, in that we can all attest that something is in this spot at this time, meaning, as in point two, the object is not context-dependent, it does not depend on our perception of it. Space is merely a container for its contents. Plus, time, matter, and energy are infinitely divisible,

implying that there is no lower limit as to how small their units can be subdivided into, nor is there an upper limit as to how fast we can travel.

   a. The world is deterministic, that is, describable as a law-abiding machine, making this a mechanistic world picture that can be described mathematically.
   b. The world is predictable, where its future state can be calculated from an accurate knowledge of its initial conditions.
   c. There are no surprises in a mechanism like this, which unfolds according to physical laws. There do appear some flaws in the system from time to time, as in any machine, like the eccentricities of the orbits of the planets, but God can be relied upon to step in and tinker with the system so as to set it on its proper path once again.

5. The world is reductionistic, in that an entirety is nothing but the sum of its parts. (Most times when you see a statement using the terms "nothing but," you have before you a reductionistic statement.) Properties residing in the part will also be found in the entirety and vice versa. And any change is just a rearrangement of the furniture, making the world a smoothly-running machine, or, to use an alternative metaphor, a clock with God as the clockmaker.

6. The world can be described using only one set of natural laws, for the large scale as well as for the small. Newton thought that this set was applicable for all times and places.

Writing more about theology than he did science, Newton judged this system as amounting to the one proper to and worthy of the divinity. Cracks, however, began to appear in the edifice. Moral of the story thus far: making assumptions can be presumptuous. Beginning with the physics, twentieth century enigmas surfaced which placed Newton's scheme in a negative light.

1. Speaking of light, experiments revealed that there are times in which light behaves like a wave and other times like a particle, namely the photon, the fundamental particle or quantum of the electromagnetic spectrum of radiation. In actuality, light is neither a wave nor a particle, it merely exhibits these behaviors. It depends on how the apparatus has been arranged: if it is set up to look for one of these behaviors, then this is what it will disclose. Thus, contrary to assumption four, context does become an issue—experimental results are context-dependent.

2. Speaking of quanta, there is no infinitely divisible continuum of time, matter, energy or infinite velocity; rather, they take on certain specific values and/or have limits. Energy, for instance, comes in discreet packets called quanta, the smallest of which cannot be further subdivided. As for time, we do not know what happened before ten to the minus forty-three seconds after the big bang since it could very well be that time itself does not occur in nature in parcels lower than this amount. Hence, contrary to assumption one, the world is not strictly as it appears since it does not immediately give the impression that it cannot be infinitely divisible—the fact that it is not fails to be immediately apparent but is borne out only through more advanced experimentation; and assumption seven, the world of the small (the microworld) is different from the world of the large (the macroworld). One cannot, as it turns out, employ one set of laws to cover both worlds; at least two are required.

3. Speaking of the microworld, it can only be addressed probabilistically—entailing that events are treated not individually but in large groups through the use of statistical procedures—which increases the probability of an event occurring at the quantum level. Examples include the wetness of water, where individual molecules are not wet until we have a large group of them; only then does the property arise. Individual water molecules are not a little bit wet. Another is table salt. Its constituents are sodium and chlorine. Well, chlorine is a poisonous gas and sodium is explosive in water, but sodium chloride is not an explosive poison. Therefore, contrary to assumption six, the world is not rigidly reductionistic, for sometimes wholes have properties which parts do not.

4. As a further illustration, take the milestone reached by Ernest Rutherford, who in 1919 discovered the atomic nucleus, where atoms are composed of positively charged protons, neutral neutrons, and encircled by negatively charged electrons. Rutherford proposed that a model could be put forward wherein an atom acts like a mini solar system as planets revolve around stars. This, as it happens, became applicable only to the simplest of all atoms, namely hydrogen, which has but one proton and one electron. There is a distinction to be made, once again, between the astronomical and quantum worlds. In the former, those planets revolving more closely around their star than outer ones need to travel faster and thus possess greater energy, otherwise they would succumb to gravity and plummet into the star. The circumstances are reversed in the latter, where electrons whirling around a nucleus are of

lower energy the closer to the nucleus they travel. When they absorb the energy of a photon, they move to a higher level or shell, and move to a lower when they release this same energy. In actuality, there is no adequate model of the atom—it cannot be imaged or pictured, since we are not able to represent or visualize probability waves. What this implies, contrary to assumption two, is that the world is sometimes incoherent and does not always make sense, and assumption four, for when the electron is likened to a cloud it connotes that simple location does not adequately describe it. It is not at a certain place at a specific time like what we would ordinarily understand as the operation of a particle. It would be more accurate to consider the electron as being spread out like butter on toast or smeared out like stains on a shirt. See, we *are* reduced to using analogies and metaphors after all. As it stands, descriptions lack precision, contrary to assumptions one and two.

5. The universe changes in time, where not only do stars form and meet their demise, sometimes explosively, but where much of the energy required for this process has been exhausted, implying that star formation has and will continue to become diminished. This contravenes assumption three.

6. Finally, cause and effect do not operate in the way we have come to expect. This is most prominently the case in radioactive substances, where we cannot point to a cause as to which nucleus will be the next to disintegrate as opposed to any other or when this will occur.[1] All of them have an equal opportunity and the playing field is leveled. Consequently, assumption five does not hold, for the world is neither always deterministic nor predictable, and it does contain surprises.

In fact, all seven assumptions topple in the twentieth century, and despite Newton's genius, he was incorrect on all seven counts. The bottom line is that we will need a more adequate strategy to invest in that takes all of these puzzles into account.

Theology 1

Proceeding to theology, Newton's successors became disgruntled with the notion of a divinity cast into the role of a cosmic mechanic or regulator who intervenes in the natural process every time, if left to itself and under its own steam, it will not yield what is in accordance with the divine intention

---

1. Barbour, *Religion*, 170.

or will. Newton retained God in his system by God's correcting the world as it stands whenever it requires it, and the form which this interference assumes is usually taken as the miraculous. Yet we must ask, as those critics who came after Newton did, what is wrong with the world that it calls for such adjustments? The proposed answers were either that the world was not made perfectly, since if it was it would not warrant such attention, thereby casting doubt on God's engineering capability, or if it was, then it is not perfect now. Neither possibility is attractive. Their recourse was to remove this responsibility on God's part by removing God from the scheme. Everything comes at a cost and this is the price to be paid. If one wants to retain God theologically, then one must remove God physically from the ongoing processes of the world.

As a recourse on the part of Newton's successors, the divinity assumed the position of only being involved in the world at its very outset by setting its initial conditions, thereby essentially winding it up and allowing it to unfold on its own as an elastic band does for a toy boat or plane. This countermove goes by the name of the aloof, detached, withdrawn, absentee landlord God of deism, who is no longer involved in the world or interacts with it. This divinity does not perform anything any longer that would make a physical difference to us. The distinction is that whereas in traditional theism God is active continually and is referred to as interventionist, in deism God was active once. The downsides of this move include the following:

1. There is no longer any use in praying to this God to change things, since God is no longer operative in the world. Prayer thus becomes ineffectual.

2. One could argue that there is virtually no difference between a deity who is no longer active in the world and there being no deity at all, for how would we recognize the difference between them?

3. Should the world unfold entirely in accordance with the way it was wound up, then all the responsibility for all the evil and suffering in the world must rest upon God. Humans are thereby divested of all this responsibility, which is beneficial on the surface, but God then becomes pernicious, which is not.

In essence, God is objectionable for either setting the stage such that violence would occur, or intervening in a seemingly arbitrary way by withholding assistance from some who could really benefit from it, as in the Holocaust, where the conventional God had the power to avoid the utter atrocity and did not but permitted it to take place. The result is that God's method of operation would be inscrutable, which at the very least, for better or for worse, is biblical, in that God's ways are higher than our ways and those ways are beyond finding out and hence inexplicable in the ordinary sense (Isa 55:8–9). Not a desirable outcome either way. The bottom line once again is that we need a more adequate strategy that can disentangle us from this knot.

Relativity

This now brings us firmly into the twentieth century and a distillation of Albert Einstein's revolutionary work. Space and time here are understood differently than at Newton's time. Despite being a disciple of Newton as we shall see, Einstein challenged his notions of space and time and was driven to announce that they are context-dependent, specifically they depend on each observer's frame of reference. For him, this affects mass, length and time. Taking each in its turn, the inertial mass of an object (its lack of changing its state of motion if left to itself) increases to infinity as the speed of light is approached. Why? Because it takes an infinite force in order to do so. Next, the length of an object contracts in the direction of motion as the speed of light is approached. Why? Because space itself contracts near the speed of light in the neighborhood of the object. And as for time, clocks run slower as the speed of light is approached and in the presence of the gravitational field of a massive object like a planet or star.

To test our comprehension of these phenomena, take the nearest spiral galaxy to our own, namely Andromeda (or M31 in the classification scheme), which is 2.4 million light years away from us. From our perspective or frame of reference, a light beam from there would take 2.4 million years to reach us, meaning that the signal it carries is 2.4 million years old, thereby entailing that its information is ancient history. We have no more recent information as to what has happened to the galaxy in the interim. But from its own perspective, the light beam makes the trip in an instant. Why? Because travelling as it does at the speed of light, its clock has stopped; no time has elapsed for it in traversing the distance. One implication of this is that photons do not age until they make contact with an object. Their clock begins ticking only when, say, captured by an electron.

Here is another test. If you are travelling at one half the speed of light, how much faster is a beam of light passing you? The answer, counterintuitively, is at the speed of light; it overtakes you as though you were stationary. Why? Because our measuring apparatus are themselves affected by the speed of light. As intimated above, rulers contract and clocks tick slower. Hence the speed of light is the same for all observers, so whenever we measure it, it yields the same result[2] up until we are at about 90 percent of it. The speed of light, then, becomes the lone absolute.

Einstein conducted a number of thought (*Gedanken* in German) experiments in his day, largely because the experimental equipment was not around at the time. At one point, armed with the above insights, Einstein contemplated that if he were to observe a clock that has a second hand and he was travelling away from it at light speed, the signal informing him that another second has elapsed never catches up to him. The kicker is that Einstein was a precocious teen when he had this insight. A contemporary example of these effects is the GPSs in our cars, which must take the above into account. The satellites affording us this information are high above the earth and its gravitational field, meaning that their clocks run faster. But they are also in rapid motion with high velocity, connoting that their clocks run slower. These do not cancel out, so the overall effect, as it happens, is that their clocks run faster and gravity in this case edges out velocity. This effect must now be factored in and corrected for, otherwise GPSs would give rise to errors which would steadily increase over time.

We thus find ourselves in the realm of special relativity, devised by Einstein in 1905. In it he tells us that mass and energy are equivalent, using his famous equation $E=mc^2$. They are the same things only in different forms, implying that a lot of energy can potentially be retrieved from a small amount of matter. Physicists have put it this way: Energy is an excited state of matter; mass is a condensed form of energy.[3] This was confirmed in the first atomic bomb blast in 1945,[4] which converted a small amount of matter into a large amount of energy. Further, space and time are not independent but are combined into spacetime, where they are distinguishable but inseparable.

Moreover, there is no privileged, or inertial, frame of reference available in the universe, that is, one which is not itself subject to motion. There is no universal hitching post or fixed reference point immune to motion. Everything moves. Even if we are stationary with respect to a point on our planet, we are still rotating around Earth's center, are still revolving around the sun, are still revolving around the center of our galaxy, are still revolving around the center of our galaxy cluster, and so forth. Dizzy yet? In essence, there is no objective time or place for anything in the cosmos; rather, putting it into simpler terms, nothing can be at rest or have zero motion, for everything and its motion is relative to everything else and its motion, and this is what is relative about relativity theory.

2. Kaku, *Hyperspace*, 83.
3. Owens, *And the Trees*, 83.
4. Barbour, *Religion*, 178.

Yet Einstein did not stop here. He concluded that this special theory applied only to motion in straight lines, at constant speed, and in empty space. But space is not empty and is curved. This drove Einstein to propose a general theory in 1915, ten years after the special. He conjectured that to be accelerated is to imagine that gravity is acting on you. If we were to be accelerated at the specific rate of 9.8 meters per second squared (m/s^2), then we would suppose that we were on the Earth's surface, since this is the rate at which objects fall toward Earth near Earth's surface. If we were to be accelerated at this rate in a windowless elevator in outer space, in another thought experiment of Einstein's, we would not be able to tell the difference between being in outer space versus on the Earth's surface. From this he concluded that the geometry or configuration of space is itself affected by the matter within it. Specifically, both space and time are distorted in the neighborhood of a massive object like the Earth. The greater the mass of such an object, the greater the spacetime curvature around that object.

Physicists put it this way: "Space tells matter how to move; matter tells space how to curve."[5] This warping of space propels the imagination to posit a force behind it, yet it is more accurate to describe it, contrary to Newton, as the bending of space; there is no "spooky action-at-a-distance" as Einstein claimed in criticism of Newton. The general theory was also confirmed, this time requiring only four years to do so, for the solar eclipse of 1919 disclosed that light rays from stars are bent slightly on the way to Earth and our observing eyes as a result of the sun's gravitational field which these rays curve around.

Once again, submitting an example to test our comprehension, given that it takes about 8.5 minutes for light to travel from the sun to the Earth, and at the risk of repetition, the signal affords us information about the sun that is in the past by being 8.5 minutes old. If the sun were to instantly blink out of existence (perhaps by the power company for failure to pay its energy consumption bills), how long would it take for Earth to fly off into space? You might be tempted to say instantaneously, but do not fall into that trap. Remember that nothing travels faster than light. Even gravitational waves take time to reach Earth. Consequently, the signal which informs you that there is no longer a sun there takes 8.5 minutes to reach Earth and you. So the correct answer is 8.5 minutes.

Gravity, as it turns out, is a property of spacetime itself, thus general relativity gives us another equivalence principle: whereas in special relativity space and time, matter and energy are distinguishable but not separable, here in general relativity gravity and acceleration are not even

---

5. Barbour, *Religion*, 179.

distinguishable. And wherever there is gravity there is spacetime geometry/curvature and vice versa. Einstein challenged Newton but was actually the last great Newtonian in the sense that he held on to the concept that everything could still be measured using rulers and clocks. This will become important when we discuss the quantum world, but for now we are ready to tackle our next exercise of theological reflection.

Theology 2

For this, we will need to recall from last time the metaphysical attributes of the classical Judeo-Christian divinity. There are nine:

1. Independence, where God requires nothing outside of God's self for existence; plus, God is self-sufficient in that God is not dependent on what God creates.

2. Spirituality, where God is incorporeal; plus, spirit does not take up space.

3. Eternality, where God is unaffected by the passage of time; plus, God is timeless.

4. Simplicity, where God is not a compound, that is, made up of parts.

5. Immutability, where God does not mutate, but remains unchanging. This is drawn from the ancient Greek philosophical view that change implies corruptibility. The reasoning is such that change is ultimately always in the direction of corruption, for material things will inevitably decay. Take humans: the end of our lives is not usually a pretty sight. Since matter changes and is ultimately corrupted, therefore God cannot change but remains static. Thus matter is devalued and spirit is highly valued.

6. Impassibility, where God has no emotional investment in the world but is indifferent, dispassionate, and unmoved towards it. This, however, overlooks or even ignores passages such as Matt 23:37 and Luke 13:34 wherein Jesus, at least, laments that he "longed to gather [Jerusalem's] children together, as a hen" gathers and collects her brood. It sounds like God does care here. Moral of the story: the Bible cannot always be used in support of theology.

7. Omnipresence, where God is everywhere present (with the possible exception of hell). There is nowhere that God is not; thus God is omniperspectival, in that God perceives from every point of view.

8. Omniscience, where God is unlimited in God's knowledge in terms of the past, present, and future; plus, God is aware not only of the way things are but the way they could be. Once again, this is a doctrine which does not find reinforcement in Scripture verses like 2 Chr 32:31b where the author(s) write about King Hezekiah: "God left him to test him and to know everything that was in his heart." The implication is that God did not already know.

9. Omnipotence, where God can fulfill all which God sets out to accomplish, with the proviso that it is not a function of power to perform the logically impossible, such as fashion a square circle.

With both relativity and these theological categories as preliminaries before us, we are now prepared to ask, if we line them up side by side and compare them, do any areas of fragility or discrepancy arise, that is, do they quarrel at some point? Is there something about one set that does not square with the other? There appear to be at least two and we can list them together with their implications.

First, there emerges an internal inconsistency between the doctrines of eternality and omnipresence when relativistic categories are applied to them.

a. If God is not in time regarding eternality, then neither can God be in space regarding omnipresence, since all there is is spacetime. One cannot separate them, for space is time and time is space according to special relativity theory.

b. If God is in space, as God is in the mythical tale of Genesis, particularly in the verse 3:8, where God walks in the Garden of Eden, or, as certain Christians would claim, is present in the person of Christ, then God must also be in time and therefore temporal, meaning affected by its passage.

c. Classical theology is insisting that we separate time and space, but relativity will not allow it.

d. Therefore there needs to be some reformulation of these metaphysical attributes as classically defined for them to survive into the twenty-first century.

Second, more speculation than outright argumentation:

a. Omnipresence dictates that God be omniperspectival, that is, in all frames of reference, which is what omnipresence means, and hence boasts a fixed or privileged frame which no one else can duplicate.

b. Independence means that God is not relative.

c. Both (a) and (b) entail that God is absolute.

d. We may ask as to whether a non-relative divinity is something that we, who only know from the perspective of a single frame of reference, can relate to?

e. Further, can such a God be relevant to those who do not share this God's lack of limitations?

f. Therefore, the classical God is in danger of not being relative, related, or relevant to us. This might be the cost or price to be paid if we want to retain (or salvage) a God who goes by the classical description, for that God does not seem to be able to operate in a relativistic universe. Once again, the bottom line is that we will need a more adequate strategy for a divinity that can last into the twenty-first century.

Given the foregoing, the new physics is not an extension of Newton; rather, Newton becomes incorporated into the new physics. The new physics contains Newton and can be used to describe Newton, but not the reverse. One cannot get the new physics out of Newton, though one can obtain Newton as a specific case from the new physics, which is the broader viewpoint.

Quantum

We now turn to the quantum world and let's commence with the electron. It is not really at any point at any time as a particle would be, meaning it is not localized, making it more like a wave. To treat it as a particle is to overlook its wave characteristics and vice versa. Yet it interacts with other particles, like photons, which are the quantum, or the elementary or fundamental unit, of light. A particle is nothing like a wave nor a wave a particle. The electron also does other odd and curious things. For instance, when it jumps from one energy level or shell to another around an atom's nucleus, which is where the term quantum leap originates, it does so without traversing the space in between. In other words, it is never part of the way or on route to another orbital, nor is there any elapsed time in the displacement.

The inquisitive among us will immediately jump all over this and object that in doing so it would contravene the Einsteinian absolute of the speed of light, the response to which is, of course it does, and proudly so. The difference is that we will need to note our whereabouts every time the observation registers. Recall that the relativistic order has jurisdiction over the macroworld and the quantum over the microworld. Whereas Newton had a one-size-fits-all approach to natural laws, here we find that each order

QUANTUM CONUNDRUM AND OTHER RIDDLES OF SCIENCE AND RELIGION   127

warrants its own set. The quantum world is not ruled by the relativistic and vice versa. So the electron does not commit any infraction nor contravene anything except our own expectations. Einstein for one had such expectations. He insisted on being armed with rulers and clocks to make measurements, but they do not assist us here, which is why he ultimately abandoned the program. His macroworld commitments prevented him from microworld vision and imagination. And even though these two sets of laws are incompatible with each other, one being nothing like the other and hence constituting grounds for divorce in a court of law, both are required for a complete description of the universe.

This is what prompted Niels Bohr and his Copenhagen school of interpretation to submit its complementarity principle, where there is a duality to nature in that entities like electrons can behave as though they were both waves and particles.[6] Plus, the two approaches, relativity and quantum, are complementary in that one covers what the other does not and without which one would not, as mentioned, be left with a thoroughgoing picture of the cosmos.

But this duality is not the only principle on offer. Werner Heisenberg then affords us the uncertainty principle, in which the more data we glean from one aspect of an elementary, fundamental particle-wave, such as the position of an electron, the less we are able to obtain about another, like its momentum. Heisenberg viewed this situation ontologically, as if to say this is the way things are and the way the world works, while Bohr perceived it epistemologically, by stating that this lets us know what can be known about the world. Here waves become waves of information and are to be understood not as water waves but, say, crime or heat waves. For Heisenberg this implies indeterminism, where the world is entirely a matter of chance; for Bohr it leads to agnosticism, since we cannot tell what is occurring in the world apart from our observation of it, which amounts to manipulating it, for the act of studying an object on the part of a subject alters it.[7] (A bumper sticker in Germany reads "Heisenberg may have slept here."[8] We just can't be certain.)

When we want to secure information about the elusive electron, we bombard it with photons which collide with it and are then reflected off it and into our awaiting equipment, the signal registering on the dials and meters. But the interaction modifies the trajectory which the electron had prior to its bombardment. This means the data secured are neither pure

6. Barbour, *Religion*, 170.
7. Barbour, *When Science*, 68.
8. Kaku, *Hyperspace*, 116.

nor objective; we do not really know what the electron was up to before we interfered with it. To Bohr's thinking, this does not describe circumstances which anticipate a time when we will have better instrumentation in practice; instead, this is a lack of knowledge in principle—a situation we cannot repair, but a law of nature, an impenetrable barrier to knowledge.[9] Whereas relativity is counterintuitive—it is not what we would expect, quantum is irrational—it does not make sense. And the irrationalism of the latter led Einstein to distance himself from quantum and part company with it even though he was instrumental in its onset.

To illustrate this interference, along comes Erwin Schrodinger who also devised a thought experiment—Einstein was not the only one to have them—which came to be called the Cat Paradox. A cat was to be sealed in a steel chamber containing all of a radioactive substance, a Geiger counter, a flask of hydrocyanic acid, and a hammer. At the end of a selected time period, if any nucleus in the radioactive substance has disintegrated, it will register in the Geiger counter, which in turn will trigger the hammer to smash the flask, releasing poisonous gas, thereby killing the cat. Admittedly, it needs to be stated at the outset that conducting an actual experiment of this sort today would not survive an ethics committee due to the potential harm inflicted upon the feline, although I suppose none can prevent us from contemplating it. So let us proceed.

The situation can be described mathematically using what is called a wave function measuring the probability of the cat's fate. What it reveals is that the cat can be understood probabilistically. While it is unobserved, there are multiple possibilities for it, hence it is referred to as suspended in a superposition of states. In essence, the system is open-ended and undecided. In order to gain information about the cat we must lift the lid, though in so doing the system becomes interfered with. Prior to this, the cat can enigmatically be described as both dead and alive or half dead and half alive, since the wave function must cover all the possibilities, paradoxical though this might be.

To retrieve data about the system's actual state of affairs, we become the observers, which amounts to interfering with it. Once the system is measured in this way, the wave function is said to collapse and yield either a dead or alive cat. This act renders the system into one of its compound states and one of the, in this case, two possibilities surfaces and becomes definite. Schrodinger used this to declare that quantum drives us to preposterous conclusions, but this absurdity has become the accepted position. He would claim that cats are only ever dead or alive, meaning this is an

---

9. Barbour, *When Science*, 74.

either-or situation, while Bohr insists that they are both-and. What has just happened can be summarized in this way:

1. Wave functions are mathematical equations describing the trajectory of a wave and are thus deterministic, as is anything reducible to equations, since plugging in certain figures on one side of the equal sign will yield only one on the other.
2. Waves describe the probability of locating a particle somewhere and are thus indeterministic.
3. As a probability, there are a range of possible outcomes, meaning the alternatives are in a mixed state.
4. Once the observer enters the stage and seeks to secure information from the system, the act of doing so interferes with it.
5. The act collapses the wave function and therefore the system into one of its component states where a definite reading can be taken.
6. This is when the wave assumes the form of a particle.
7. It is not known how or when the transition occurs and what constitutes the observer, which is not always obvious; in our example it could be the person who lifts the lid, or the Geiger counter, or even the cat if it lives.

The above yields the long-standing debate between Einstein and Bohr concerning what and when something can be known in the quantum world. Whereas the former argued that what is real can be measured—that it already possess properties even prior to measurement, the latter counters with the assertion that the measurable becomes the real, implying that the act of measurement confers these same properties on it. Einstein would argue that one can speak of properties before observation; Bohr submits the counter-proposal that they surface only upon observation.

Insisting that his view was the correct one, Einstein teamed up with two other physicists named Podolsky and Rosen in 1935 to concoct another thought experiment termed, you guessed it, the EPR experiment. They proposed that when a source emits two particles initially in contact with each other and are hence referred to as entangled, and then fly off in opposite directions, they must have opposite spin, for conservation laws require that that the total spin be zero.[10] If you measure the spin of one, therefore, you already know the spin of the other even without measuring it, for it must bear the opposite spin. Buoyed by this ammunition, the team announced

10. Barbour, *When Science*, 82.

that, as a result of this line of thinking, one can know properties even prior to measurement.

Bohr, undeterred by this challenge, responded that if one can achieve this, then regardless of how far apart the two particles travel, they will eventually become too distant for the speed of light to be the signal between them. Consequently, as he confidently reported like someone who has just checkmated his opponent in a chess match, knowledge then *can* occur faster than the speed of light, contrary to this absolute. At that point Einstein could only retreat like a vanquished foe, dismayed that his thinking had let him down by not anticipating this objection, into the defensive posture of uttering that there must then be some rational explanation for this, that something must have been overlooked, and that there need to be hidden variables to account for it. But none was ever found.[11]

What can be concluded about this phenomenon is that particles of this kind continue to influence each other until such time as they are interacted with. They too are in a mixed state awaiting observation or measurement. This is contrary to Einstein's entrenched stance that there is no probability here and the particles must already possess definite values. Bohr adamantly differs in stating that what we have before us is not two separate particles but a two-particle system, or, one system with two parts, in which the speed of light is not a consideration. This effect is what is termed non-local—occurring more quickly than the speed of light can facilitate. It is important to be aware that this is the quantum world and there is little that is ordinary about it, given that we are used to living in the macroworld of relativity. While the jury is still out on this issue, the physics community has mostly followed Bohr in it, meaning that even Einstein has his critics.

Entanglement is therefore required for non-locality but not the reverse, on the grounds that entangled particles need not be distant from each other. There is, however, a proviso. Admittedly, observing a system disturbs what we are attempting to investigate; but this is unable to account for what obtains when nothing disturbs the system, as in the unpredictability of which will be the next nucleus to disintegrate in a radioactive substance. We cannot point to a specific cause responsible for the decay of one nucleus as opposed to any other. In essence, what would be the unidentifiable cause of the selection process behind the unidentifiable cause of the actual selection?

At this stage we can provisionally conclude that whereas physics has been touted as an exact science, at the quantum level it lacks precision and it conducts its commerce with the economics of indeterminism and the currency of probability. Moreover, Einstein has earned hero status, due

---

11. Barbour, *When Science*, 83.

largely, I suspect, to his media-assisted image of an eccentric uncle. If we are looking for modern-day myths, Uncle Sam is one, for he is a fiction—there isn't one; and Einstein is another, since he is heralded as being the greatest mind of all time. Bohr, his peer and equal in certain respects, gets much less press outside his native Denmark, and not many other than those within the physics community have ever heard of Podolsky or Rosen.

As a summary and series of asides, together with introducing some philosophical concepts into the mix, the following is offered. To begin with, Newton was incorrect in that one set of laws can apply to the entire cosmos. Quantum has undermined this view. We now require two sets. For the macroworld, we refer to Einstein, through whose work we have a sense of the nature of the universe. But this is where it ends. For the microworld, a complementary-to-yet-incompatible-with-relativity set of laws assists us in navigating our way through the small scale. Newton thereby is not so much supplanted as he is placed in a corner as a valuable preliminary case.

For most of our everyday world of experience, such as small masses at low velocities, Newton is adequate—the world can be explained as nicely Newtonian. Yet as long as the universe is inhabited by elementary particles, the constituents of all others, like quarks (the constituents of protons and neutrons) and electrons, neither Newton nor Einstein will account for all of it—we will require quantum for the remainder. (Though keep in mind that, despite being fundamental, not all particles, like neutrinos and photons, are building blocks, for they do not go into the making of much of matter as necessary ingredients.) And this is where the difficulties commence, for relativity and quantum are incompatible—not even as straightforward as comparing apples and oranges, but more like fumbling with a Rubik's Cube that has no solution.

Additionally, something we have not broached as yet, fields have virtual particles which transmit their force: the electromagnetic force is conveyed by photons; the gravitational by, what else?, gluons, and so forth, that pass rapidly between objects in their field. And as intimated, waves are best represented not by those we find in water but, dare I say it, a wave of communicable diseases, thinking medically (sound familiar?), you know, the virus that looks like a minuscule version of dryer balls or Ferrero Rocher chocolates.

Before the Standard Model of physics undergoes significant renovation, which it might in due course what with the findings about the mystifying muon, I thought it expedient to mention where we currently stand. Do we think we are not tradition-bound? We still confer doctorates even in, say, chemistry, on candidates who successfully complete and pass their rite of passage, known traditionally as the oral defense, and call them PhDs (Doctors of Philosophy), despite their perhaps never having stepped into a philosophy course in their lives. The reason? Just like Tevye in "Fiddler on the Roof" would say: tradition! This stems from medieval times and the onset of universities.

Also recall that equations are deterministic: plug in values on one side and you will get only one answer on the other. So it is important to remember that this also occurs to an extent in the microworld, since it is called quantum mechanics after all. The wave function is a deterministic equation, we simply do not always know which values to expect, the reason being that we get in the way of our own observations of that world.

Relativity might tell us about some aspect of reality, while quantum can tell us mostly about knowledge—an epistemological category. And whereas the relativistic world is counterintuitive, the quantum is irrational, but this has not propelled science to curtail its endeavors in them. Quantum events do not cease to be real, but much of the reality is in the knowledge we can glean from them. A quantum object has information to yield, but we have access only to part of it—the part we interfere with. The remainder is genuinely unknowable and we are left in the dark.

Recalling an example, the next nucleus to decay/disintegrate in a radioactive substance has no well-defined cause and as such is genuinely unknowable. Our knowledge extends only to what we elect to measure through the apparatus we have set up. The quantum objects are real both before and after a wave function collapse. Like the radioactive substance, there is a finite number of outcomes that a quantum event can yield. Which one is only ever an issue of probability. We might have an interest in a specific outcome, but the reality is that we cannot determine or enact it, and knowledge is restricted to a post-probabilistic event. Quantum events do not extricate themselves from the level of information. To suggest that objective reality is the currency quantum trades in to confuse categories. It is not in the business of divulging a reality that we at our level have encountered before. Rather it tells us what the odds are that we can eke certain information out of it.[12]

Once information is retrieved, we are then limited by our language constructs in order to convey it. We observe from our own personal perspective and interpret the experience of the characteristics of an event or object prior to encountering it. Our mentality constructs reality and the world is pre-interpreted even before we meet up with it, thus nothing in perception is given. Neuroscience corroborates this. Our brains have the tendency, nay habit, of filling in the blanks of our vision. We usually do not notice our ocular blind spot, for instance, unless it is pointed out to us. Plus,

---

12. Barbour, *Religion*, 172.

our brains work with a model of the world built up from our experience of it on the basis of our expectations, and it requires some effort to revise them when needed. As such there is no pure objectivity available to us. This is not new. We can only ever experience the world with the equipment with which we have been bestowed. The reality is the event; the knowledge is our experience of it, and our perception yields information which we report on using language, which itself is dependent upon the shaping influence of its historical context.

There are, understandably, difficulties at each stage. We come to an event with baggage. This imposes a grid upon our experience. We attempt to conceptualize the experience and situate it in terms with which we are familiar. After forming concepts we translate it into language to convey it to others. We compromise on accuracy at each step, creating the kind of difficulty the biblical prophets confronted when attempting to capture what was for them an inexpressibly moving religious experience. We lack the wherewithal to do justice to the event at each phase. Despite having a clear vision of human nature, not even God sees objectively, since even God elects to view the world through the lens of what the Messiah has accomplished for the world. This amounts to a choice on God's part, and it does distort God's vision, fortunately for us.

To reiterate, relativistic features such as distances becoming shorter the faster we travel in that direction (good to know if we are late for work), our (inertial) mass increasing with increasing speed (so why bother jogging to get in shape?), and our clocks run slower with greater velocity (great if you want the moment to linger), and these are counter-intuitive (though at terrestrial speeds the differences are not appreciable). This is not what we would expect.

The situation is different, however, with quantum. Permit me to submit another paradox. The ancient Greek philosopher Zeno offered us one with which we still wrestle. He presented us with the perplexing circumstance of arrows shot from a bow that must first travel half the distance to its target, then another half, then another, then still another, to the point where it can never reach its target since there are an infinite number of halves to traverse. The resolution, of course, is that infinite divisibility does not entail infinite distance, so the arrow will unfailingly arrive at its intended destination (the intention on the part of the shooter, that is, not the arrow's itself, provided, of course, that s/he has sufficiently good aim). Careful if you attempt this at home. Zeno marks one point where rationalism fails us and empiricism rescues us, for lo and behold, runners both begin and end races after all and tortoises lose them. In the quantum world, once again contrary to Newton, there is no infinite divisibility, for, as intimated, energy levels come in specific units, lower than the lowest of which we cannot go. There is, in effect, a physical stop sign at the lowest level, one which electrons, at least, do not contravene.

Plainly, quantum transcends the simply counter-intuitive and makes it look insipid by comparison. That world is confounding. Awaiting a response from publishing houses is analogous to the Cat Paradox. Prospective authors are faced with a superposition of states on the part of each and every publishing house whose editors are in a decision-making process mode, entailing that the manuscript proposal anticipates the collapse of the wave function into one of, in this case, its two component states—either accepting or declining (or let's call it what it really is, a rejection, even though it hurts deep inside to use the term). The observer can be either the editor or the author of the piece awaiting in eager expectation of the dreaded email informing him, her, or them of the editorial decision. Once the email is either drafted or opened, the collapse of the wave function occurs and might be accompanied by the collapse of the author-observer. Given the odds of getting a manuscript published, the probability is usually in favor of the decline component. A pox on the paradox!

Theology 3

An exchange between a theist and an atheist could proceed as follows: "Why do you believe in God? It is irrational." "Well, so is quantum, but that does not stop the world from believing in it." The question to be posed, though, is are they irrational in the same way? To claim that the latter has the backing of science is insufficient. In like manner, to assert that God's operation through both determinism and free will as merely a paradox to accept is unsatisfactory. Some elementary particles have a tendency of making an appearance when the apparatus is set up to look for them. This makes one wonder if, in the extreme, one could do the same for additional particles that one suspects might show up given the right conditions. If accurate, would that then be a product of setting up the proper circumstances or the sheer willing them into existence?

Thus far, this is dissimilar to purporting God to exist. There are definitely those who would psychologize the notion and insist that one's imagination is the deciding factor in conjuring God up from a concept to a being, thereby ontologizing something abstract. At minimum, it could be maintained that a parallel obtains between the two pursuits in that a certain amount of imagination is factorial in conceiving God and previously unexperienced particles. Yet one can live one's life without ever giving God a first thought, never mind a second one; and there are those who place a considerable amount of cerebral effort into keeping the idea of God at bay. But for them the irrational part of it stems from the perceived faulty indicators as to what would themselves constitute a divine presence.

Anything that can be pointed to, from their perspective, which would for some proportion of the world convince them that God is a reality: from the miraculous to amazing coincidences converging to result in our arrival on the scene so as to debate the very question, to the operation of conscience, to what makes us separate from the other animals (less than we realize), to love for another as a reflection of the love of God, to life in an orderly cosmos, all having rational explanations, leaves the irrational as the remainder. The ordinary senses which science utilizes are of no assistance, so runs the argument, in establishing the existence of the divinity, making God non-empirical—a decided disadvantage for those who rely on nothing else. This of course assumes that our usual five senses exhaust the ones we come with; any other, like intuition, is probably just private, not public—another downside as seen by the scientific community, even though scientists come with and use it too, as does law enforcement with its hunches.

This differs from the irrationality of the quantum world, as the confounding aspects of it belong to its behavior, not its contents. Despite no one ever having isolated a quark or captured a neutrino, electrons are beyond doubt even with their fuzziness and amorphous cloud-like qualities. Dark matter and dark energy may be hypothetical entities, but not so with electrons. The behavior of dark matter and energy, though, can be mystifying. The wording here is intentional. There are those who have "faith" in the existence of these hypothetical entities and insist that they must be real, since how they are believed to act explains the operation of galaxies, for instance, and without which we would be at a loss for an explanation, so there is a lot riding on them.

Is this any different from some individuals contending that, say, transformed lives are the result of God's interventionist influence in their lives? Only, perhaps, in that a non-material propensity, like hatred, were we to nurse it, is met by another non-material impulse, like a spirit. In the scientific realm, at least entities are physical objects. This makes the scientific angle more credible for some. However, we can also surmise that hatred is irrational. Hence, it seems that the two types of irrationality differ qualitatively (not knowing how one would quantify either of them), meaning they are not of the same kind. This implies that the notion of a person resting content with the belief that God exists is deserving of credence, since quantum constitutes the same, is wide of the mark. The comparison is not at the same level, though this does not affirm that the scientific side has greater or superior confirmation. Instead, the view that there is room for having confidence in both could be considered a rational one. From my lights, and I rarely disclose my own cards, there is a sense in me in addition to the typical five which such a spirit is able to connect with. For this I am grateful.

We are now in a position to compare the quantum world with the attributes of the classical divinity. What, for instance, would the deity be up against with quantum and which attribute(s) would be in jeopardy? Unpredictability, for one, could very well cripple God's ability to know the future as an element of omniscience. This would depend on whether the situation is in principle or in practice for God. The possibilities could be framed in the manner below:

1. If by indeterminism we mean unpredictability, then can God know which will be the next nucleus to decay in a radioactive substance?

   a. If so, then the event must have an identifiable cause after all, which could be the divine in the role of Supreme Observer or Grand Collapser of wave functions. God would then have knowledge by interfering with the world in an interventionist capacity. God would know the future by making the future, implying that God's knowledge is in practice. Knowledge then becomes derivative of intentional activity, but we must ask if this really counts when merely making the future gives God the knowledge of it.

   b. If not, then God forfeits omniscience of the future.

2. If by indeterminism we mean uncertainty, then does this mean that God can observe without interfering?

   a. If so, then knowledge once again is in practice for God, since God could presumably overcome what for us is a barrier. God would then know undistorted information, and observing would not mean distorting.

   b. If not, then neither is God omniscient concerning the present, for even that would be unknowable in principle for God until, like us, God elects to interact with the system.

These are not knock-down arguments, nor are they intended to be, but they do raise suspicion and even cast doubt upon the attribute of omniscience, at least as classically formulated. What compounds matters is that quantum outcomes cannot even be selected, that is, one cannot choose a definite component state from among the available ones, since even that is a question of probability. There is no guarantee that the desired result is the one that the collapse of the system will yield. One might wish for, say, an alive kitty, but one cannot enact it. The issue is whether the outcome is in doubt also for God, and, if so, in practice or in principle? If the latter, then God would have difficulty acting in a definite way at the quantum scale were this to be God's intent.

The future would be open even for God, who would need to gamble on outcomes, making it a risky business. God must then reckon with chance, and this is the nature of the process divinity outlined last time: for whom omniscience is confined to the past and present and does not regard this ability not extending into the future as threatening but is resigned to persuading entities to craft a future for themselves, to the extent that they can, in accord with the divine purview of the world; where self-determining power is a genuine capacity for humans directed toward an open future; and where God is comfortable with the unavoidability of chance. If God is to make a difference to us, if God is to affect the world in ways that we would notice, then God must take the world God has been instrumental in drawing out into account, like its relativistic and quantum aspects, for this is the one we inhabit. Failure to do so would entail that God's activity would go unnoticed.

As I have presented previously, one can have an open future without adopting the process portrayal wholesale. The openness divinity of free will theism resides in a position on an axis somewhere in between the extremes of classical and process. At least we know that we are not left without options were we to decline the future aspect of omniscience.

Others

Three topics remain for us to address. The first is cosmology, which treats issues surrounding the beginning and end of the universe. There are several models for this. Only one occurs for the beginning, since the big bang model has no serious competitor, nor is it probable that a contender will be forthcoming. Not as such, however, for the end.

On the one hand, a closed universe describes a scenario in which the cosmos contains sufficient matter for gravity to force a universal re-collapse, where the big bang will have a limit to universal expansion, which will slow down, arrive at a maximum point, and then snap back like an elastic band. The big bang would then yield a big crunch. An open universe, on the other hand, would find itself bearing insufficient matter to force this re-collapse and instead will expand until such time as there will no longer be enough kinetic energy remaining so as to perform work. Whereas the closed universe will end in a heat death, the open one would yield a frozen death; fire or ice are the alternatives.

Yet these are not the only ones. A flat universe would also continue to expand without re-collapse, though it would do so at a decreased rate, requiring more time to arrive at a similar end as the open model. The final possibility comes in two parts. An oscillating universe would produce a new big bang subsequent to each big crunch, thereby eliciting next generation universes. In essence, the universe would bounce and this will result in a series of bangs and crunches without ceasing. The other type is similar in that there would not be bounces of equal amplitude, so to speak, as the previous one each time a new universe is given birth but would only reach ever-decreasing extents, entailing that the era of universe-production will eventually come to a close and end once again as the open one is expected to.

The one which will likely obtain is the open form, for the main concern is the value of what is known as the cosmological constant, which is a measure of the repulsive force or anti-gravitational effect of space itself leading to its expansion rate. This value has been refined since January and March of 1998, which is when the articles describing its approximation were published.[13] They revealed that we live in a flat universe[14] in which its expansion is not only proceeding but accelerating,[15] meaning it amounts to a runaway expansion,[16] resulting ultimately in an open universe in overdrive. If humans survive long enough, we will observe the effect by witnessing the disappearance of galaxies which currently reside at what for us is the edge of our visible universe, as we in our galaxy would do the same for observers there. And this will continue more rapidly as time progresses. Should the divinity elect not to bring history to a halt prior to this, then that will describe the fate of our cosmos.

With the above information in hand, we are driven to ask the question as to whether it seems likely that this is the kind of universal ending scenario that a divinity would orchestrate. Of the available alternatives, is it surprising that the divinity would opt for the open model as the one for the end of the universe if a different one could have been arranged? Does this one best serve the divine intention and is it the one most befitting a deity? While we might not be thrust into the role of God's press secretary, one may be hard pressed to imagine that an at least ultimately open universe does not

---

13. Wilford, "New Data," "New Astronomers."
14. Gates, *Einstein's*, 209.
15. Croswell, *Universe*, 221.
16. Croswell, *Universe*, 224.

result in a public relations nightmare for those who are compelled to ask, "Is this the best your God can do?"

All of this, of course, is unless the universe does something completely unexpected as it did about five billion years ago. Inexplicably, at that juncture the universal expansion began to accelerate. And since there is no known reason why it did so, nor can it be known, if we are honest, we also do not know whether a similar event in reverse, however unlikely, is to come.

Another question which could be raised in this connection is whether humans can be considered significant in such a vast cosmos. And on the heels of this question arises another issue known as the anthropic principle, anthropos being Greek for human and from which we get the name of the discipline termed anthropology. This line of inquiry involves the staggeringly precise features of the universe that must be in place were life to appear and then give rise to readers of this work. There are three versions referred to, in order, as the weak (WAP), strong (SAP), and final (FAP) types, speculation increasing with each successive one in its turn. The WAP states, among other things, that the universe provides or has been provided with the physical constants which can produce not only life but creatures with mind. So far this is minimally controversial, since we are already here to reflect on the events that brought us here so as to enable us to reflect upon them. The ability to reflect on them entails that they in fact were in place.

The SAP elicits a more conjectural position such that, once again in hindsight, the universe possessed an aim, end, goal, or purpose in bringing about consciousness in some form in order that the universe could observe itself. There was no other way available than the way things turned out. Lastly, the FAP is the stuff of artificial intelligence (AI), wherein information processing by some intelligence will never disappear once it appears, hence once here, always here. The difference is that the bearer of this capacity need not be biological. The reader will note that we have not as yet made any reference to the divine, though some are driven to assert that God was behind it all, directing it, navigating it, or choreographing the dance.[17]

There is also a competing view to the anthropic principle which suggests that there are many universes, not surprisingly called the many-universes hypothesis, so many in fact that one of them would inevitably be the one bringing consciousness-bearers about, and we are the ones fortunate enough to find ourselves in it. This approach, employing as it does the currency of chance, also need make no reference to a deity. We are then left with the question as to whether the universe is designed to bring us forth and, if

---

17. Barbour, *Religion*, 242.

so, does it imply a Designer? In any case, chance need not operate as a deity-deterrent as will be demonstrated below in the next section on evolution.

Jumping headlong into the fray, the stages of the evolutionary process can be summarized as follows:

1. Limited resources (and all of them are, there are no infinite resources) impinge upon organisms and may call for adjustments on their part;
2. which leads to competition and the struggle for life;
3. which highlights variations;
4. which leads to adaptation of traits and selection for usefulness in an environment;
5. which leads to fitness, survival value, and success rate—the leaving of more offspring than one's competitor;
6. with accumulation of adaptive variations, this leads to a new species;
7. and with time these organisms can no longer interbreed with their parental stock.

While ordinarily not so straightforward or commonplace, what this means is that differences in degree (variations) can in fact lead to differences in kind (species).

Chance is usually understood as being in opposition, that is, a threat to design and hence to God. There are, however, three theological responses to chance, with an initial statement that neither chance nor time is operative as such, rather they are passive, not active. They do not act but are the arenas within which action occurs. The first riposte is that God is an active agent involved in events, but there resides a perception problem here. God may indeed execute a predetermined plan, but we might perceive it as a series of random occurrences. The divinity's incognito activity then becomes attributed to the very chance we imagine is antithetical to God.

Second, God actually incorporates chance into an otherwise lawful system, implying that God is the God of chance as well. In order to be genuine, though, this scheme must amount to risky business on God's part where something at least has the potential of catching God by surprise. And whereas the first position could be viewed as an outworking of the classical theistic strategy of those like Thomas Aquinas, the second, as mentioned, would describe the approach endorsed by the openness model, otherwise known as free will theism. The third strategy maximizes chance wherein God lacks control, since entities enjoy self-determining power, and so is thrust into the role of acting by way of influence or persuasion alone, and this describes the process perspective.

Charting the differences, God might be active in position one, authentic chance may be found in position two, but only in position three do we find both. In the latter alternative, the process deity recognizes its own limitations and can therefore not guarantee best outcomes, since entities have a decisive role in the shape results will take, for better or for worse. Questions which evolution can spark include whether the direction natural selection takes necessitates a Director; in addition to whether it is more dignified or ennobling to be specially created or to emerge through a long and messy random history; and if natural selection is something we should marshal our efforts against. Further issues which to date have thwarted the wisdom of the discipline include where the entrance of slime mold is to occur in the classification scheme, since it seems to arise abruptly, together with what would be the ancestral forms of species like the octopus and squid, for they do not appear to have differed since their inception, although admittedly invertebrates without exoskeletons do not customarily leave behind much in the way of fossils.

These questions and others like them are ones I placed on the final exams for the students in my religion and science classes. But fear not, nor despair, for I gave them the questions beforehand so they could prepare and asked them to answer only three. The exercise for the entire term was to ponder if the classical divinity could survive into the twenty-first century, or instead whether a revolution is warranted in theology as there have been in science. They found it liberating even to ask the questions regardless of where their answers led them.

IS IT TIME TO PANIC YET?

In an earlier treatment, I addressed the issue of whether God would bring history to a close once the world becomes subject to more extreme droughts, famines, plagues, and so forth. Not least of which is the pernicious problem of fertility—male sperm count and motility have alarmingly become reduced (some of these concerns being traceable to what we ingest in our food, such as particulate matter from soft plastics (called micro-plastics) in food containers, and we can even absorb them from cosmetic containers), and infertility is dramatically on the rise for women—implying that our species' longevity is in peril regardless of any external dangers.

Nevertheless, several of these types of disasters have and are occurring on a regional and not a global scale, but the pandemic example has definitely become global. By virtue (or lack of it) of our international travel together with insufficient personal and social hygiene and environmental devastation practices, as researchers lament, once the virus is somewhere it reaches everywhere. Once our population numbers, unfettered travel, and deforestation policies exposing new viruses increase, so do incidences of such concern. What were once isolated problems are not isolated anymore, but transcend geographic and political boundaries, and as such are more difficult to contain, since our increased numbers entail that we are too close to each other so as to avoid them, unless we are very careful and work in conjunction to end the threats.

From a theological perspective, we can ask, as others did when having encountered previous disasters, as to whether God has left us to our own devices in order that we may witness how our complacent actions can and do recoil on us. The scriptures teach that God has too much invested in the world and so will not abandon it to chaotic powers (John 3:16; Rev 21). There are times, however, when the optics seem to differ. These questions are not confined to our situation at present, but have been asked on several other occasions, and when they were, at least some persons believed they were facing the beginning of the end.

Taking a few examples in reverse chronological order, the Bay of Pigs in the early 1960s between the U.S. and Cuba together with, at the time, the Soviet Union could readily have elicited such a reaction; so too with the scourge of the Third Reich; and again with the influenza outbreak, the appellation Spanish being a misnomer, at the end of World War One. And while we are on the topic, WWI itself, the innovation of a global battle, reveals that new does not always translate into the improved. Thankfully, these events did not usher in the end of the world, despite how incendiary they became. So is the recent crisis any different?

Another threat to the human race is the discovery that male sperm count, at least in the West, has declined by the alarming rate of one-half in the time span of a one-term president, and it shows no signs of recovery but continues its downturn. One causal factor, as mentioned above, is the presence of hormone inhibitors ingested from micro-plastics in the diet. Female fertility has also taken a hit. The cumulative effect can most assuredly make us feel like God is absent. We tend to yearn for conditions reverting to normal, though God perhaps disapproves of our version of it. New routines replace former ones and the world proceeds apace without further consideration. Maybe they are opportunities to reflect, but they come at too dire a cost. Humans have the propensity and tendency to seek to avoid engaging with the world as though we were numb to it or sleepwalking in it. Disasters upset this equilibrium and we seek to tame the offense so as to return to the comfortably dull. As time goes on, regrettably, threats become more difficult to repel, and hurdles and roadblocks to our security appear to elude eradication.

We might pacify ourselves by thinking that the world came out the other end with the previous crises, so why should this one prove resistant? "We will get through this" may become our personal mantra; we might even believe it if repeated often enough. We have not come across a foe that we could not overcome; will this be one of them? The only one, or the first of many? Microbes could mutate at a rate we might never get ahead of. If not this one, then maybe one to come. We cannot tell, no matter how well we prepare. Our presumption could be our undoing.

We are given a glimpse in the book of Revelation as to what the end will look like, specifically there will be a new heaven and earth (chapter 21). The description, though, must make the Earth substantially different from what it is now. The New Jerusalem is symbolically purported to be a cube measuring roughly fourteen hundred miles on each side and with walls two hundred feet thick (21:16–17). In comparison, Mount Everest is about as large as a mountain can become on Earth without damaging its crust, and it is still growing at a rate of approximately two inches per year, since the Indian subcontinent tectonic plate has not ceased to slam into Asia. As a result, the Earth's crust could not withstand a city of this magnitude, hence we must be speaking of a significantly modified Earth or resign ourselves to the idea that the biblical description must be symbolic.

Revelation informs us *that* the end of the age will arrive; when and how we are not told, at least not definitively. Will it be prompted by certain disastrous events, or will the perfectly ordinary lead up to it? Revelation itself seems to inspire doomsday scenarios, so the former usually gets the nod in the popular imagination. Some persons and/or organizations believe they have it all figured out, but as the Jesus of the text cautions us, not even he knows (Matt 24:36; Mark 13:32). In particular, he employs the non-menacing metaphor of the fig tree and the turning of the seasons to summer when the tree bursts into bloom as figurative for when the time is nigh; thus he knows the season, but nothing more precisely than this (Mark 13:28–37). All we can be certain of is that we are closer to it today than we were yesterday. Most significantly, for those interested in exoplanets, in the previous two verses, Jesus announces that he will "gather his elect from the four winds, from the ends of the earth to the ends of the heavens." Looking forward to discovering who these might be. Best then to prepare not only for the end of the age but for a multitude of alien introductions.

CONCLUDING

Several of the foregoing segment titles have ended in the form of a question, even though this is not the game show *Jeopardy*. The reason is that there are often more questions than there are answers, or once there are answers, they elicit more questions. This state of affairs is not likely to diminish, at least in the short term.

Allow me to speak of truth again, as in the last volume. Truth has not as yet died out, or at least specific truths. One is that we all bear subjectivity. Hands up those who are without it. I thought so. In this sense, subjectivity is objective in that we all lay claim to it. That is about as objective as objectivity gets, for there is no uninterpreted reality or actuality. Pure observation and objectivity are beyond our reach. Another truth is that we are not, contrary to the military commercials, all that we can be, nor can the army make us so. This is because there is always both room for improvement and no one who is unassailed by regrets. Still another truth is the most controversial one, that a way through our character blemishes or tarnish has been accomplished and provided for us, we not being up to the task. This is as much a hope as it is a purported truth.

The biblical text informs us that nothing is impossible for God (Matt 19:26). That, however, is clearly not the case. In the history of philosophy, there have been, and still are, some metaphysicians who have argued that God must contain the world, which means the cosmos, more accurately all universes, implying that there can be nothing other than or distinct from God, for to their thinking it would limit God if something were to be external to God. Whether or not God contains the world, though, is not the issue, because God is still limited, since not even God, for instance, can change the past. Our perspective on it can most definitely change, yet this is not changing the past but only our present response to it. This also occurs for God. For those in God's program, God sees our past through the lens of the Messiah's reparation. Hence God wears corrective lenses which distort God's vision so as to view the world through the work of the Messiah, such that our rough edges and missteps are no longer held against us and we are re-shapen to reflect our Shaper. Thus Jesus' achievement serves God as ultimate optometrist (since one of God's appellations is the Ancient of Days (Dan 7:9), and we know what happens to sight as we age).

Consider the modified messages we are given from the OT to the NT. In the former, the Law or Torah is regarded as God's greatest gift to humans, since it informs us as to how to live in a way most pleasing to God and toward other humans, and it is seen as a benefit and a joy, not a burden, to apply to one's life. But, as mentioned above, take 1 Kgs 19:15–21, where the prophet Elijah (18:22, 36) is commanded by the God of the text to "anoint Elisha . . . to succeed you as prophet." During the exchange, Elisha requests that he be allowed to return to render his farewell to his parents. Elijah consents to it. Then in the NT, the Jesus of the text is faced with a similar situation as with Elisha. In Luke 9:61–62, he is asked by a would-be follower if he could bid his parents farewell and then follow Jesus, but is scolded, "No one who puts his hand to the plow and looks back is fit for service in the kingdom of God."

An opposing OT passage worthy of attention is the destruction of Sodom and Gomorrah. Lot and his wife and two daughters were prompted to flee before this sentence could be carried out and not to look back (Gen 19:16–17, 26), though, regrettably, "Lot's wife looked back and became a pillar of salt." Those reading or listening to Luke's gospel, if some of them were Jews, would have had these two OT passages in mind. Yes, they would have acceded, one ought not look back, yet a major figure as Elisha was given leave to do as he requested, and if he could "look back," then why not a prospective follower of Jesus? He had the tendency of beginning some of the teaching accredited to him with the formula, "You have heard it said, . . . But I tell you . . ." (Matt 5:21 and many others) and in so doing was emphasizing the Spirit in addition to the spirit of the law. Elisha's method of operation is no longer deemed acceptable behavior for admission into the kingdom. In comparison with formal education, the OT way would be analogous to an introductory course, whereas the way of Jesus in the NT would be the advanced approach. Or, the OT would be for the newcomer, the neophyte; the NT for the more developed, as it comes at a greater cost.

But allow us to conclude with three sections devoted to our primary concern in these pages, namely religion, science, and the two combined, respectively.

On the religion side, it is one thing to differ, even vigorously so, with the considered opinions of others, assuming they are in fact duly considered and not simply salvos from the hip; it is quite another to sit in judgment on another's character or spiritual status. Should the object of one's opprobrium be the Judeo-Christian sacred text or the theology which is drawn from it, the temerity on the part of the investigator can be evaluated as off-limits on the part of those invested in biblical and theological standing, with the objective that these pursuits not dishonor the tradition. In their estimation, what is seen as casting aspersions upon either the Bible or its theological derivative, together with church councils, creeds and confessions, originates from the enemy camp and needs to be squashed.

This mentality, though, depends on one's commitments and whether one presupposes that the Bible and theology are human or divine products, or both, for to find fault with either of them stems, it is believed, from the apostate at best or the demonic at worst. Should one's commitments lie in the sphere of divine product, the believer feels pressed into service to defend the text from antagonistic assaults. The trouble is that this heartfelt allegiance toward the protection of the sacred is misplaced. Some detractors are under the mistaken assumption that theology is as authoritative, to the extent that it is in fact authoritative, as the sacred text from which it is generated but at best can only approximate. Theology, however, cannot nor should it have the same status as the scriptures, since it is derivative of what is itself not above or beyond criticism. The danger is to idolize theology such that God must conform to our systematic conception of God.

The problem can be framed in this way: whereas the OT text is a product of many human hands over many centuries from an Hebraic worldview in an oriental historical context, the NT Common Era texts owe their provenance to a Greek worldview and often an occidental setting over the course of about a single century. Concomitantly, the very quest to systematize the statements in the text is an occidental (Western) and not an oriental (Eastern) pursuit. The latter in time, namely theology, is an exercise about which the Hebrews displayed little concern, they being more inclined to focus on wisdom and endeavoring to live their lives accordingly. The Greeks, on the contrary, despite the term philosophy translating into the love of wisdom, were all about coherence of propositions and the love of reason or rationality. Under their admittedly skillful hands they attempted to massage the statements (read "lop off the extraneous bits that did not fit") into shape such that they would not be party to contradiction, because, after all, there is a law against that sort of thing. The efforts which comprise this project reveal that systematic theology is very much a human undertaking.

What this means is that theology can only ever approximate the thrust of the biblical message. Theologies come and go into and out of favor and the Hebrews had little appetite for their construction. For them, it requires wisdom to navigate through the apparent inconsistencies and assess the circumstances under which each is appropriate, since there is a time for each situation (Eccl 3), despite the opposition that the statements themselves might exhibit on the surface (is there really a time to hate?). This is the way of wisdom, not insisting that the text conform to an imposed grid other than the internal correspondence to the law. Rather, the sentiments found in the text demonstrate a human attempt to come to grips with sometimes inexpressible religious experiences. The divine, it would seem, supplies the experiences; humans provide the statements to capture them and the program to commit these experiences to writing and bequeath them to future generations; the latter two proving imprecise.

This is where the opposing theological camps come to loggerheads, and this language would be perplexing to those invested in the text, so much so that they feel they must defend the scriptures, the battle lines having been drawn. Yet this is bibliolatry, the worship of the Bible itself, veneration which should be reserved for the divinity alone. Nor is it an unexpected stretch that this mindset is also applied to theology, let's call it theolatry, especially when the longstanding tradition involving such hall of fame heroes as Augustine and Aquinas is called into question. Consequently, the troops see fit to unleash an onslaught against the gainsaying theological infidels. Anything deviating from this is deemed heretical and can no longer remain under the banner of Christianity, for those who engage in drifting from the time-honored tradition have deserted the faith and cannot be counted among the faithful.

Anyone having a different perspective on the faith or vision of divinity is vilified, since these views sound so foreign to what the orthodox are used to and hence the latter feel justifiably threatened by them without ever engaging with them. This mentality prevents them from recognizing the value of considering alternative model proposals so as to offer the tradition what some might regard as what is missing, namely much needed refinements, for we should not shy away from what may be closer to an approximation of what the message was intended to convey, and about which God is not threatened. As a litmus test, have we thrown our hats into the theological ring to the point where we think it cannot be improved upon? But if it can, then why stand in the way? Otherwise former alliances tend to be dissolved and previous friends become foes.

Instincts on the part of the church are not always reliable when they differ from body to body, each under the impression that they alone are the ones truly led by the Spirit. Given this situation, can each community definitively claim that the Spirit is speaking particularly to them when the messages themselves can vary so widely? The Spirit, we have come to believe, does not act chaotically but orderly. Should the tradition-bound assert, however, that they simply follow God, they shamelessly surmise that theological dissenters do not, and they then go back to their work in creating division and building fences in place of bridges. Some of them feel it is their job to assume the worst and are well adept at it. A pestilence on their presumption. All we are left to do is to pray for their judgmentality. And should you suspect that I speak from experience, you would probably be correct. Investigation should not be forestalled.

Additionally, Christianity tends to stack the deck in its favor. There is no knowing what post-death experiences will hold without going through it oneself, and even then we are informed that there is no knowledge in the grave (Eccl 9:10) but only sleep. Besides, calling up spirits from the dead to communicate this information to us is strictly prohibited (Lev 19:31). As a result, the church can fall into the trap of feeling it has a monopoly on the afterlife's GPS coordinates, with the assurance that we, with divine assistance, can get there from here.

The sacred text gives directions but only offers some clues. We are not aware if afterlife events will play out as described in its accounts, and no amount of research will unravel or settle the issue. So we might be driven to trust in the ecclesiastical and scriptural messages, the two not necessarily identical. This can lead to an abuse of power where people can be taken advantage of and the vulnerable exploited. There is an unavoidable risk involved, and if we are detectives attempting to solve this mystery, we will not get very far, for we are kept in the dark. Despite the danger of mixing metaphors, the divine economy has us over a barrel and each of us individually must decide on the course of action we are to take, the received or inherited tradition not always being entirely of assistance. Hopefully we can summon the courage to embark upon the quest in a way that promotes inclusivity, for shunning is not becoming for the church.

On the scientific end, presuppositions surface there as well. Here are a few of the philosophical problems associated with it. Science operates with the empirical method, which incorporates the usual fivefold sense-oriented experiences and observations regarded as the most reliable route to knowledge. Now it is time to apply this very same methodology toward science itself. So what do we observe? First, the move from objects in our senses, affording us knowledge of them, to suggesting their ontological existence in reality is illegitimate, for what we observe does not necessarily conform to or confirm what in point of fact obtains. Anyone having suffered from hallucinations, I say suffered for there are rarely beneficial ones, can attest to this. As such, we can only report on our observations, otherwise this amounts to a metaphysical assumption that our senses provide us with the existence of something independent of them, and this debate shows no signs of subsiding.

Second, physics and chemistry were hardly behaving as exact sciences when they posited the phlogiston theory, where combustion releases this hypothetical substance, thereby making up for the decrease in weight of the remnants. Moreover, the idea that the universe is infinite can hardly be the product of experience. It is a reasonable enough assumption, but this does not get us beyond it and into the realm of observation. Additionally, there was also the impression that light must travel through a medium in space, for there must be something substantially continuous from the source to the receiving object, since this is how matter works, so let's call it the ether. Neither phlogiston nor the ether ever received confirmation. Furthermore, the case of the disastrous effects that thalidomide inflicted upon progeny in the 1960s should also serve as a caution that medical science can get it gloriously inaccurate as well. Generally speaking, technological advancements are not always immediate improvements until such time as they can be safely tested in the public sphere. This makes science and technology ambiguous in that they can be either a boon or a bane. Time will tell in each case.

There has been demonstrably much less to convey in this section than the one above, but it is no less weighty or significant for that. In both this and the previous one the undertakings can also definitely benefit from refinements, and learning from our mistakes is one of them.

Lastly, we examine the relationship of religion and science and whether and how one affects the other. In order to have a consistent and thoroughgoing amalgamation of the two, one must have a shaping influence on the other, but what would be the likelihood of this prospect and what would it look like? Well, one can invite science to inform the contours of one's faith, particularly when it comes to, say, the findings of archaeology together with what paleo-anthropology tells us about human origins, our early history, and our evolutionary development.

Should the former disclose that the number of those having perished in OT battles has been inflated, owing to the dearth of physical evidence about it—there ought to be skeletons and weaponry in abundance if the recorded events actually occurred as stated, and this is unsupported—then one must accept that the figures are hyperbolic and that the documents are somewhat less than reliable. This would then shed negative light on their trustworthiness as historical accounts and prompt us to suggest that perhaps they had a more mythical or propagandistic import to yield. And should the latter maintain that there is an unbroken line of descent from proto-humans to humans instead of a Golden Age of humans specially created and untainted by disobedience, that there is no indication of a living soul having been imparted to already existing humans at any stage in history, and that there is no evidence of a time on earth when a curse was visited upon it, then this must have a shaping influence on our theology.

CONCLUDING 151

Having said this, science must also be willing to discuss what the sparking of our pre-frontal cortex into action, of a type not previously enabled, at a point in time approximately one hundred thousand years ago, what a Great Awakening of humans roughly between fifty thousand to forty thousand years ago launching us into a religious consciousness, among other things, and what specific major religions having been initiated within a few centuries of each other, referred to as the Axial period, without having come in contact with each other so as to compare notes, what effect these could also have on theology, potentially proposing as they might more traditional categories of divine activity.

Hence the influence is not unidirectional. Science can also be undertaken religiously as in the case of the anthropic principle. This type of examination need neither lead to nor stem from a religious perspective, but it can. Such investigation can prompt some individuals to conclude that a divinity can lie behind the amazing coincidences of the universal history leading to us. This approach is not for everyone, but it does highlight one strategy for those intending to make the relation a two-way street. Science would then point to potential parameters within which faith can be placed (in addition to the preoccupation with alliteration on the part of some).

Religion can also provide parameters for science, at least for believers, perhaps in terms of gratefulness toward the deity's providence in establishing an orderly world and thereby permitting scientific study into it which then furnishes us with law-like generalizations reflective of this order. This much may be granted, though religion can further play a more substantial part in the research into the fine-tuning of the cosmos. Plus, if we insist that there is nothing that is not natural, if nothing is permitted to count as the divinity attempting to pierce through to our interiority in ways that are not based on science, then these divine messages might not be heard. In placing such a grid on our world, God does not have a voice and so must act through nature. We have forced God's hand and not given God any other choice, and so we are left with God's response to "reap the whirlwind" (Hos 8:7a). There is continuity here with an ancient pre-scientific mindset.

For science to be truly science, in the spirit of discovery, it must leave the door to the possibility of nature's divine provenance open, or at least must not close itself off from extramundane imagination by hanging a shingle on the door reading "No Divinity Allowed."

I debated with myself, since there was no one else in earshot at the time, as to how best to bring this volume to a close. It became clear to me that when all else fails, simply turn on the news for a spate of relevant material. Sure enough, this I did, and where I learned that Canada, at least at the time of writing, has just passed a bill into law banning the use of what is known as conversion therapy. It has been applied in the past (which is what the past tense terms "has been" imply, thus making the combination "has been" and "past" redundantly superfluous) to rescue members from alleged cultic organizations, more precisely as a type of deprograming intervention, a practice which some consider laudable. The new law does not extend to this but to the "de-alternate lifestyling" of someone, which is not laudable, as if to say that through coaching and discipline one can be rid of the dreadful trans, gay, or lesbian blight.

At best this can only serve to make people reject themselves, who they really are, by using methods geared to breaking down a person's self-esteem and urging them to accept that they are reprobates, thereby instilling in them the need to decide to join the converting group, for therein lies esteem and freedom from reprobation (Reprobateness? Reprobatehood? Reprobatitude?), something which certain cultic organizations are allegedly familiar with and as such *should* be outlawed. This could potentially, now that I think about it, also affect the Church of Scientology for their alleged questionable psychological practices (on the advice of legal counsel, I liberally employ the term "alleged").

Science might perceive these circumstances as not something which darkens their own precincts, but not so fast. That may have been accurate before, but not so much recently. The oft-heard call to heed and believe in the science is not what science has had to concern itself with, namely an advertising and marketing strategy. Yet here we are. Science informs us about climate change, the microbial threat, and the benefit of vaccination, though not everyone is convinced. It seems that science is attempting to convert the unwashed unbelievers over to its side after all. One more way in which both science and religion have similar boats to paddle. If we carry the comparison further, we could even submit that at least biology has its own sacred text, particularly Charles Darwin's *Origin of Species*. Physics and chemistry, on the contrary, do not appear to have such a time-honored script. So there you have it, one way to combine religion and science in the same sentence. Consider this my attempt to convert some in the either-or camp into a both-and approach, trusting that it will not become another target of the ban.

Finis / The End

APPENDIX

The following Scripture verses are those employed in the chapter "The Bible: Edited Version." Note that those passages which are parenthetical in nature though not surrounded by parentheses (a total of five) are placed in square brackets.

Genesis: 2:12; 13:10b; 14:2b, 3b, 7, 8, 17b; 19:22b; 23:2, 19b; 25: 30b; 30:35; 35:19, 27; 36:1; 46:8, 12.

Exodus: 4:26; 9:31–32; 11:3; 15:23b; 16:36; 30:23; 38:22–23.

Numbers: 12:3; 13:16, 20, 22b, 33; 26:29, 33, 46, 58b–61.

Deuteronomy: 1:2; 3:9, 11, 13b, 16, 17, 19; 10:6–9; 13:6b–7, 13b; 14:29.

Joshua: 3:16; 9:1b; 12:3, 7b–8, 9, 23; 13:27b, 31; 15:9, 10, 13b, 15, 25, 36, 49, 54, 60; 16:2; 17:8, 11b; 18:13, 14, 28; 19:2, 8, 47; 21:10, 11(2), 13, 21, 27, 32, 34, 38.

Judges: 1:10, 11, 23; 7:1; 8:24b; 9:18; 13:16b; 14:4; 19:10, 16; 20:27–28.

Ruth: 4:7.

1 Samuel: 9:9; 14:18; 20:39; 23:6; 27:8b.

2 Samuel: 1:18b; 9:10b; 11:[1], 4b; 21:2, 12b.

1 Kings: 1:6; 4:10, 11, 15, 19; 7:47; 9:16–17, 20; 10:11, 12; 11:2, 13:18b, 18:3–4; 21:25–26; 22:38.

2 Kings 15:37.

1 Chronicles: 1:27; 4:18, 22b; 5:1–2, 23, 26; 6:10, 54, 57, 67, 77; 8:12; 12:1–2, 19; 13:6; 26:10.

2 Chronicles: 4:13; 8:7; 9:10–11; 20:2b.

Ezra: 2:1–2, 6, 16, 36, 40; 3: 8, 9, 10; 4:11; 10:19, 23.

Nehemiah: 7:6b–7, 11, 21, 39, 43; 13:2.

Esther: 2:15; 3:7.

Proverbs: 7:11–12.

Isaiah: 28:[23–29]; 29:10(2); 66:19.

154 APPENDIX

Jeremiah: 2:11; 29:2.

Ezekiel: 45:14b; 47:16.

Daniel: 4:8b, 19; 7:12.

Jonah: 1:10b.

Mark: 3:16, 17; 5:41; 7:3-4, 19; 9:6; 10:30; 11:32; 15:16, 42.

Luke: 1:70; 6:14; 9:14, 33b; 23:19.

John: 1:38, 41b, 42b; 4:9b, 25, 44; 6:16; 18:5, 10, [14]; 19:13b; 20:9, 24; 21:7, [19], 20b, [23].

Acts: 1:15, 18-19, 23; 2:11; 4:36b; 9:36; 11:28b; 17:21; 21:29.

Romans: 2:14-15; 10:6, 7.

1 Corinthians: 1:16; 7:10, 12; 9:20b, 21b.

2 Corinthians: 11:23.

Ephesians: 2:11b; 5:9.

Colossians: 4:10b.

Hebrews: 10:8b.

Revelation: 19:8b; 20:5.

BIBLIOGRAPHY

Aczel, Amir D. *The Jesuit and the Skull: Teilhard de Chardin, Evolution, and the Search for Peking Man.* New York: Riverhead, 2007.
Barbour, Ian G. *Myths, Models, and Paradigms: A Comparative Study in Science and Religion.* New York: Harper & Row, 1974.
———. *Religion and Science: Historical and Contemporary Issues.* San Francisco: HarperSanFrancisco, 1997.
———. *When Science Meets Religion: Enemies, Strangers, or Partners?* San Francisco: HarperSanFrancisco, 2000.
Barnett, Lincoln. *The Universe and Dr. Einstein.* rev. ed. New York: Perennial Library, 1966.
Behe, Michael J. *The Edge of Evolution: The Search for the Limits of Darwinism.* Toronto: Free, 2008.
Bohm, David. *Unfolding Meaning: A Weekend of Dialogue with David Bohm.* New York: Routledge, 1987.
———. *Wholeness and the Implicate Order.* New York: Routledge, 1983.
Bohm, David and F. David Peat. *Science, Order, and Creativity.* Toronto: Bantam, 1987.
Briggs, John P. and F. David Peat. *Looking Glass Universe: The Emerging Science of Wholeness.* New York: Touchstone, 1986.
———. *Turbulent Mirror: An Illustrated Guide to Chaos Theory and the Science of Wholeness.* New York: Perennial, 1990.
Broad, C.D. *Lectures on Psychical Research.* International Library of Philosophy and Scientific Method. New York: Humanities, 1962.
Croswell, Ken. *The Universe at Midnight: Observations Illuminating the Cosmos.* Toronto: Free, 2001.
De Lubac, Henri, S.J. *The Religion of Teilhard de Chardin.* Translated by Rene Hague. New York: Image, 1968.
Ehrlich, Paul R. *Human Natures: Genes, Cultures, and the Human Prospect.* New York: Penguin, 2002.
Gates, Evalyn. *Einstein's Telescope: The Hunt for Dark Matter and Dark Energy in the Universe.* New York: Norton, 2009.
Giberson, Karl W. *Saving Darwin: How to be a Christian and Believe in Evolution.* New York: HarperOne, 2008.
Gilkey, Langdon. *Creationism on Trial: Evolution and God at Little Rock.* San Francisco: Harper & Row, 1985.
Fagan, Brain. *Cro-Magnon: How the Ice Age Gave Birth to the First Modern Humans.* New York: Bloomsbury, 2010.
Frieman, Joshua A., et al. "Dark Energy and the Accelerating Universe." *Annual Review of Astronomy and Astrophysics* 46 (2008) 385–432.
Funk and Wagnalls Standard Desk Dictionary. New York: Funk & Wagnalls, 1980.

Griffin, David Ray. *God and Religion in the Postmodern World: Essays in Postmodern Theology*. SUNY Series in Constructive Postmodern Thought. Albany: State University of New York Press, 1989.

———. ed. *Physics and the Ultimate Significance of Time*. Albany: State University of New York Press, 1986.

———. ed. *The Reenchantment of Science: Postmodern Proposals*. SUNY Series in Constructive Postmodern Thought. Albany: State University of New York Press, 1988.

Haught, John F. *God and the New Atheism: A Critical Response to Dawkins, Harris, and Hitchens*. Louisville, KY: Westminster John Knox, 2008.

Hedges, Chris. *When Atheism Becomes Religion: America's New Fundamentalists*. Toronto: Free, 2009.

Hitchens, Christopher. *God is not Great: How Religion Poisons Everything*. New York: Twelve, 2009.

Holmes, Arthur F. *Ethics: Approaching Moral Decisions*. 2nd ed. Contours of Christian Philosophy. Downers Grove, IL: InterVarsity, 2007.

Kaku, Michio. *Hyperspace: A Scientific Odyssey Through Parallel Universes, Time Warps, and the Tenth Dimension*. Toronto: Anchor, 1995.

Kurzke, Hermann. *Thomas Mann: Life as a Work of Art: A Biography*. Translated by Leslie Willson. Princeton, NJ: Princeton University Press, 2002.

Lovelock, James. *The Ages of Gaia: A Biography of Our Living Earth*. The Commonwealth Book Fund. Toronto: Bantam, 1990.

———. *Gaia: A New Look at Life on Earth*. Oxford: Oxford University Press, 1982.

Lukas, Mary and Ellen Lukas. *Teilhard: The Man, the Priest, and the Scientist*. New York: Doubleday, 1977.

Mann, Thomas. *Death in Venice: And Seven Other Stories*. Translated by H.T. Lowe-Porter. New York: Vintage, 1954.

———. *Doctor Faustus: The Life of the German Composer Adrian Leverkuhn as Told by a Friend*. Translated by John E. Woods. New York: Vintage, 1999.

———. *The Magic Mountain*. Translated by John E. Woods. London: Folio Society, 2000.

Mariani, Mike. "American Exorcisms." *The Atlantic*, December 2018. https://www.theatlantic.com/magazine/archive/2018/12/catholic-exorcisms-on-the-rise/573943/.

Miller, Kenneth R. *Finding Darwin's God: A Scientist's Search for Common Ground Between God and Evolution*. New York: Harper Perennial, 2002.

Owens, Virginia Stem. *And the Trees Clap Their Hands: Faith, Perception, and the New Physics*. Grand Rapids: Eerdmans, 1983.

Peat, F. David. *Infinite Potential: The Life and Times of David Bohm*. Reading, MA: Addison & Wesley, 1997.

Primack, Joel R. and Nancy Ellen Abrams. *The View From the Center of the Universe: Discovering our Extraordinary Place in the Cosmos*. Toronto: Riverhead, 2006.

Randall, Lisa. *Dark Matter and the Dinosaurs: The Astounding Interconnectedness of the Universe*. New York: Ecco, 2015.

Rozin, Paul and Edward B. Royzman. "Negativity Bias, Negativity Dominance, and Contagion." *Personal and Social Psychology Review* 5.4 (2001) 296–320.

Simon, Charlie May. *Faith Has Need of All the Truth: A Life of Pierre Teilhard de Chardin*. Toronto: Clark, Irwin, 1974.

Stein, George. "The case of King Saul: Did he have recurrent unipolar depression or bipolar affective disorder?" *British Journal of Psychiatry* 198.3 (2011) 212.
Stringer, Chris. *Lone Survivors: How We Came to be the Only Humans on Earth*. New York: Times, 2012.
Strong's Exhaustive Concordance of the Bible. Carol Stream, IL: Hendrickson, 1988.
Sullivan, Walter. *We Are Not Alone: The Search for Intelligent Life on Other Worlds*. rev. ed. New York: Signet, 1966.
Suzuki, David. "The Equalizer." *The Nature of Things*, CBC TV, February 6, 2022.
Teilhard de Chardin, Pierre. *Building the Earth*. New York: Discus, 1969.
———. *The Future of Man*. Translated by Norman Denny. New York: Harper, 1969.
———. *Hymn of the Universe*. New York: Fontana, 1970.
———. *Man's Place in Nature: The Human Zoological Group*. Translated by Rene Hague. New York: Fontana, 1971.
———. *Le Milieu Divin: An Essay for the Interior Life*. New York: Collins, 1962.
———. *The Phenomenon of Man*. New York: Fontana, 1965.
Thomas, Lewis. *The Lives of a Cell: Notes of a Biology Watcher*. Toronto: Bantam, 1975.
———. *The Medusa and the Snail: More Notes of a Biology Watcher*. Toronto: Bantam, 1980.
———. *The Youngest Science: Notes of a Medicine-Watcher*. Toronto: Bantam, 1984.
Throckmorton, Burton H., Jr. *Gospel Parallels: A Comparison of the Synoptic Gospels*. 5th ed. Nashville, TN: Thomas Nelson, 1992.
Wade, Nicholas. *Before the Dawn: Recovering the Lost History of Our Ancestors*. New York: Penguin, 2006.
———. *The Faith Instinct: How Religion Evolved and Why it Endures*. Toronto: Penguin, 2009.
Ward, Keith. *Is Religion Dangerous?* Grand Rapids: Eerdmans, 2006.
Wilde, Oscar. *Collected Works of Oscar Wilde*. Ware, UK: Wordsworth, 1997.
Wildiers, N.M. *An Introduction to Teilhard de Chardin*. Translated by Hubert Hoskins. New York: Harper & Row, 1968.
Wilford, John Noble. "New Data Suggest Universe Will Expand Forever." *New York Times*, January 9, 1998. https://www.nytimes.com/1998/01/09/us/new-data-suggest-universe-will-expand-forever.html.
———. "Wary Astronomers Ponder an Accelerating Universe." *New York Times*, March 3, 1998. http://www.nytimes.com/1998/03/03/science/wary-astronomers-ponder-an-accelerating-universe.html.

www.ingramcontent.com/pod-product-compliance
Lightning Source LLC
Chambersburg PA
CBHW071429160426
43195CB00013B/1849